AN OUTLINE OF PHYSICS

An Outline of Physics

By

L. Southerns

[1920]

First Published in 1920

CONTENTS

PART II

APPENDICES

PREFACE

IT is necessary, in the case of a book planned in such an unfamiliar form as this, that a somewhat full explanation of the principles on which it has been written should be offered to the teacher or student into whose hands it may fall.

The business of the teacher of "First-year Physics" seems to me to comprise three things. The first and chief of these, without which, indeed, the others are hardly worth while, is to inspire in the student a keenness for careful and scientific thought. Next in importance is instruction in the actual subject-matter of Physics. And the last is the teaching of matter useful for the subsequent career of the student—to-day, especially, this cannot be entirely ignored.

But this classification by no means represents the view-point of the average student of this subject. Physics is a compulsory subject in many courses of study, and the majority of students take it up for that reason. Any value it may have in their eyes depends on its utility in relation to their principal subjects—Engineering, Medicine, or what not. Only a minority take the subject for its own sake. If what follows appears at first sight to leave out of consideration this minority, it must not be assumed that I have little regard for them. But they present no special problem.

At the outset, then, the fact of a strong divergence between the view-point of the teacher and that of the student must be faced. It is quite a natural if an unfortunate one, and is specially pronounced in this subject of Physics. To ignore it is fatal.

Now the order of the three features of the teacher's work which I have specified above is reversed in the student's view. I think it necessary to make a deliberate attempt to raise the view-point of the student, not by long range attraction, but by meeting him in the first place on his own ground. In fact, his interest must be gained—not a merely superficial interest, but a vital one—and I believe that this can be done only in one way: that is, by showing him (to his own satisfaction, not merely *telling* him) what is surely the case, that the subject is one which has a definite utility with respect to the principal subject of his studies.

Now the fact that Physics forms part of the preliminary course in many different branches of study is due to the fact that it really has an important bearing on those branches. It is one of the main roots from which they spring. Applications and illustrations of its principles abound in all these subjects. It is necessary to use illustrations of these principles continually in any course of lectures on Physics. But an illustration chosen from the student's own subject will not only be more effective than one not so chosen so far as its immediate purpose is concerned, but will also impress the student with the utility of Physics, and so help to accomplish the purpose the importance of which I have emphasized above. I must, however, here point out that the objects of an illustration of the kind I have mentioned need to be kept clearly in view, otherwise the teacher may easily find himself falling into the mistake of teaching technical rather than pure Physics—a thing to be carefully avoided.

These considerations suggest the advisability of dividing the students into classes according to the different subjects of their after-studies, and this is emphasized by the fact that different groups of students, say Engineers and Medicals, look on our subject from very different angles. Not only will the illustrations which are suitable for one group be unsuitable for the other, as is evident from what has gone before, but for the best results a certain difference of treatment of the whole subject is necessary for the different groups. Hence I believe a subdivision of the "year" on these lines to be essential. The extent of this subdivision cannot of course be prescribed here.

Having in mind ideas such as these, it is clear that I could not attempt to write a text-book which should comprise a complete course of Physics for all first-year students, so far at any rate as the theoretical part of the subject is concerned. This book is intended to be used in conjunction with a theoretical text-book, selected with special relation to the particular student or group of students concerned. Why, then, write this book at all ?—that is, the theoretical portion of it ; for as regards the practical part of the course I think that somewhat different conditions apply. This brings me to another aspect of my apology.

Physics, decidedly, is not a popular subject. Unpopularity may be partly, or even largely, due to the method of its presentation. An uninspiring course may easily do more harm than good. Against a fragmentary knowledge of a few facts and principles acquired may have to be debited a lifelong aversion from the subject.

The first year should not be a year of "getting over the drudgery." It should be a year of inspiration. How, then, should the subject be presented? Not after the manner adopted, necessarily and rightly, in a systematic treatise. Broad outlines should first be presented in an elementary but scientific manner. The subject should not be divided into water-tight compartments. A text-book necessarily divides up the subject in this way, and completes the treatment of one section before commencing the next. But this does not correspond to the natural way of gaining knowledge. Almost as well might one try to understand a picture by examining it carefully a square inch at a time, beginning at the left-hand top corner. I am not criticizing text-books, but I think that the text-book method is not the true teaching method. The text-book is an indispensable adjunct to the course. Part I of this book aims at giving a general sketch or outline of the subject, which is intended to act as a kind of frame on which detailed instruction may be hung. The outline should be given first, and the detailed instruction afterwards. The text-book from which the latter is drawn should, as I have already said, be adapted to the particular requirements of the students.

Another point which seems to me to be of great importance is that new knowledge in the subject should not be avoided. Nor should it be added as a sort of afterthought, or as a reward for attending to the other, and drier, parts of the work. It should be incorporated in the course itself, or its scientific value will be diminished or lost. A theory is not necessarily more difficult to understand because it has been developed within the last decade or two.

I have already spoken of a classification of students depending on their varying relations to the subject in general. Another classification, which cuts across this, might be made with reference to the preliminary knowledge of the subject with which the students are equipped when they enter upon the course. This classification is one of the greatest importance; but multiplication of classes soon reaches the limit of practicability. The following scheme forms an attempt to deal with this matter in a satisfactory yet not impracticable manner. Suppose A and B to denote the categories into which the students are to be divided, and a and b the grades —a representing the more advanced section, and b the beginners. We have then the four sets of students, Aa, Ab, Ba, Bb. Consider Aa and Ab, who belong to the same general order. These require the same kind of teaching, differing from that suitable to the B's ; but more elementary instruction is needed by Ab than by Aa. Suppose, now, the lecture periods to be represented by 1, 2, 3, etc. Probably the early lectures or lessons could conveniently be given to the whole class (Aa, Ab). After a certain number of periods, say 9 for purposes of illustration, the course proper begins. Lecture 10 forms an elementary introduction to a certain topic. This is developed in lectures, say 11 and 12. Now for lesson 13 the class is divided into its Aa and Ab sections. These sections may meet concurrently or successively as may be most convenient. The lessons should be informal, and will require in general little or no experimental preparation. Section Aa will pursue the topic of lectures 11 and 12 to a further stage of advancement, or study it in more detail than was necessary in the main course. Section Ab, on the other hand, will receive elementary preparation for the work to be considered in lectures 14 and 15—preparation which can be dispensed with in the case of section Aa. In this way, the boring of Aa with elementary work in which they have already received a sufficient grounding is avoided, and the time saved enables students to obtain a somewhat more advanced knowledge of the subject. On the other hand, the elementary difficulties of the beginners can be much more readily dealt with in the sectional than they could be in a general class. Lectures 14 and 15 will then be attended by the whole A class, and so on.

In this scheme the elementary, or Ab, work would constitute the normal course, on which the examination would be set. Admission to the advanced section would be open only to such students as were able to give definite evidence of their ability to profit by it. It could not advantageously be compulsory, but should be confined to those who wished to enter. Many students on entering college have a fairly good knowledge of one or more branches of Physics ; in such cases the student might be classified as Aa in these branches, and as Ab in the remainder. It must be fully understood by all students that the more advanced work is not required for examination purposes. But an examination should never be looked on as a mechanism for holding back students who are capable and desirous of going a little beyond its requirements.

A double subdivision of the complete year is thus accomplished without serious multiplication of classes. The rigidity of a single course system, which seems to me to be incompatible with really satisfactory teaching, gives place to a high degree of flexibility. Part I of this book, if used for these A and B classes during the early part of the session, need not be looked on as supplying a hard and fast course. A judicious omission here and there, in the case of the elementary sections, will do no harm. Experimental illustrations and examples, likely to interest—in the sense already explained—and instruct, should be freely given.

With regard to laboratory work, Part II of this book comprises a course suitable for general purposes. No rigid programme suitable for all students can be given. The plan here adopted is one which allows the greatest elasticity and scope for modification.

Now, a laboratory course may easily be an exceedingly dull affair, consisting of a set of discrete and unrelated, and often, to the student's mind, almost meaningless and useless measurements. Single unrelated experiments should seldom or never be given. Never—unless some definite and useful object is to be accomplished thereby. Each experiment should have a setting which exhibits its utility or scientific purpose. It must never degenerate into a mere laboratory task, to be undertaken for the sake of a mark. Many measurements which, picked arbitrarily from a list, give little evidence of their value, and are usually performed as mere tasks, would be seen in quite a new light by the student were they put in their proper places in a scientifically arranged scheme of work.

An Example is given in Part II of a useful plan, which we may call the group experiment method. Here a group of students combine to make a survey of several properties of a substance, such as common salt in solution. The method can be varied—the case given is no more than a sample—from session to session. Of course it forms but a small portion of the year's work for the students taking part in it.

Another set of experiments, not intended to be worked on the group system, is given under the heading Vibration. Many topics of this kind can be chosen by the teacher, and connected courses of experiments devised to suit. In the present case the measurement of the density of air finds a natural place. In this setting the measurement has an actual relation to the subject of experiment. It is at least as useful as if it were made as an isolated determination in an orthodox course, but here it has also a definite scientific purpose and an added interest to the student. This is not a mere question of making the work more pleasant. It is of the highest importance that the scientific bearing of any piece of work undertaken should be grasped, otherwise little advantage will be gained by its performance. In my own experimental work I do not make isolated measurements which are not required in the course of the work in hand. I see no reason for doing so, and I do not know why a student, perhaps not possessing my interest in the subject, should look at the matter differently.

In arranging a series of experiments such as we are now considering, there is a certain advantage in including items from different branches of Physics, so as to show the inter-relation of these branches. In this vibration section experiments might well be included on the simple and compound pendulum and on moments of inertia. In fact, the examples of Part II might frequently be amplified by the inclusion of such mechanical experiments as have direct bearing upon them. Most of these would be omitted by students attending a separate mechanical laboratory. A few mechanical experiments are described in a section of Part II.

A useful piece of work, which should be carried out during the course of the session by each student, is the preparation of a thesis on some selected topic. In suitable cases, two or more students might collaborate for this purpose. A number of subjects for these essays should be proposed for the class—though if some of the members were capable of choosing, and interested enough to choose subjects for themselves, so much the better. Suppose the Vibration set to be taken as the basis of one of these essays. A connected account of the elementary theory of the subject, together with a full description of the various experimental verifications and other determinations, each in its logical place, should be given. The value of such an exercise, conceived and carried out in a scientific manner, could hardly be over-estimated, however simple the plan of the work might be. Great scope exists here for grading the work to the capacities of individual students. In no case should too elaborate a scheme be tried: the value lies not in quantity, but in the scientific and logical quality of the work.

The remainder of Part II deals with various subjects, and covers a course suitable in general for the year's work. But no rigid and permanent scheme should be adhered to. While the logical connexion of the several experiments in one of these groups should be especially kept in mind, it may not always be practicable to work through them in the order given. Sometimes an experiment must be postponed owing to the corresponding part of the theoretical course not having been reached. But, even so, a great balance of advantage will remain due to the adoption of the methods suggested. It is important that the student should read the whole of a section before commencing to work the experiments in it, so as to obtain a coherent idea of the value of each part in its relation to the whole.

A certain amount of constructional work is valuable. The making of a simple sextant, and fitting up of a galvanometer from parts supplied by the laboratory, are given as examples. Many other instruments, simpler than these if desired, could be made from time to time by various students.

An illustration is also given, as a suggestion, of the kind of problem for calculation which has a definite bearing on the laboratory work. This principle should always be kept in mind when examples are being set and liberal use made of it.

Students often fail to get real working ideas of the things with which they deal in the laboratory. For example, the ability to judge, however roughly, the electrical resistance of a piece of wire at sight is often lacking. If measurements of resistance have been made only with boxed up coils of wire, this, of course, is not surprising. Such tasks are suitable only for examination purposes. The measurement of something enclosed in a box is not a very scientific proceeding, unless the student has himself put it there, or at least has a definite knowledge of what the box contains. In the Electricity section of Part II several resistances of wires are proposed for determination. Samples of these, of different materials and thicknesses, might be cut—all having the same resistance—and fixed to a card, with names, lengths, and diameters noted upon it. Simple devices of this kind easily repay the small amount of time spent upon them.

My thanks are due to my colleague, Dr. R. W. Lawson, for the assistance he has given me by reading the proofs.

L. SOUTHERNS

PART I

AN OUTLINE OF PHYSICS

PART I

CHAPTER I

The Material Universe—Physics—Inertia—Mass—Law of Force—Law of Reaction—Acceleration—Unit of Force—Physical Dimensions.

The Material Universe

THE material universe offers abundant scope for studies of many kinds, and we have been equipped with wonderful, yet not complete or perfect, means of pursuing them. Our senses, by the aid of their proper organs, enable us to perceive certain aspects of the outer world; but we must not suppose that things actually exist just as we perceive them. A musical instrument, we say, emits sound. But sound, in this sense, only means a peculiar tremor of the atmosphere. The sensation we experience as we listen is our own interpretation of this tremor, after it has been duly modified by our sense organs and conveyed to the brain. Nothing at all resembling this sensation really exists apart from the consciousness of the listener. So inanimate nature has been said to be eternally silent. Our colour sensations also are merely our interpretations of the various light-waves which reach our eyes from the objects around us. These objects are not "coloured" in the sense of possessing anything similar to the sensations we experience on viewing them.

It is very difficult to understand why objects are seen in their true positions external to ourselves. Actually, we have only two little inverted pictures of them on the retinas of our eyes—the observed scene is the result of our interpretation of these. This could not be made by a person who had the sense of vision alone. The child has to learn, by actual failures in his attempts to grasp it, that the moon is a long way off. It is only by co-ordinating vision with other senses, and by moving about among the objects seen, that we obtain our final visual impression of the external world. This important part of our education is accomplished in early childhood. It seems to give us a true idea of the actual objects around us, and in this respect differs from the other cases to which reference has been made.

But, if we wish to understand things as they are, it is clear that mere observations made by our senses will not be enough. We shall have to compare, and discriminate, and reason. We shall have to frame hypotheses and invent experiments by which these can be tested. In fact, we shall have to bring into use another part of our natural equipment—the purely mental or intellectual part. Again, instruments must be invented, which, like the telescope and microscope, will enable us to extend our sense-observations. The telescope has revealed multitudes of stars which would have remained unknown to us without it. But our knowledge of the existence of any stars depends, first, on the fact that we possess the sense of sight and, second, on the extraordinary transparency of the atmosphere. Our senses are few—there may be whole aspects of nature of which we are quite ignorant on account of the lack of appropriate senses, or of other conditions necessary for their perception. But, though our apprehension of nature may be fragmentary, it is extensive and varied enough to form a sufficient basis for our studies.

Physics

Our subject of Physics is only one of those which have for their purpose the study of the material universe. It is a wide and fundamental one. I shall not attempt at present to give a definition of it. Indeed, I think that definitions are often almost meaningless to those who have not already a pretty good idea of the thing defined.

When studying material substances, and the influences which they exert on each other in various circumstances, we find that a few root principles apply to them all. But our investigations soon reveal differences. Occasionally new phenomena present themselves which would be startling if observed for the first time by a person old enough to begin the study of Physics. A man familiar with the warmth produced by rubbing two sticks together, but who had never seen fire, might well be surprised at the result of striking a match. Many people to-day would be quite astonished at the effect of throwing a little water on a fragment of sodium.

One of the universal properties of matter is Inertia. A body at rest does not of its own accord start into motion. If in motion it does not of its own accord change its motion.

Inertia

A billiard ball at rest on the table remains so till put into motion by some cause external to itself, and when in motion its tendency is to continue to move uniformly. It would do so if it were not prevented. We can set the resting ball in motion by striking it with a cue or with another ball, and can deflect or bring the moving one to rest by the same means. By the constant use of our muscles in moving all kinds of objects we get an idea of the kind of effort which it is necessary to apply to a body in order to alter its state of rest or of motion. To this effort, as applied to the body (not to the muscular sensation which we feel to be associated with it), we must give a name. This name is Force. We really do not understand exactly what takes place when one body, by striking or pushing another, sets it in motion, but the effect is the same as that produced when we apply muscular effort to the body. But we call the process "application of force," and hope that some day we may learn more of its actual nature. We may sum up by saying that "Every body perseveres in its state of rest or of uniform motion in a straight line, except in so far as it is compelled by forces to change that state." This, the "law of inertia," is the first of Newton's* three laws of motion. The law deals with both inertia and force. We must deal with these quantitatively. It will be simpler to consider, instead of the impulsive forces of cue on ball, etc., a case in which a steady force is applied continuously to a body. Imagine a frictionless but heavy railway wagon standing on a pair of horizontal rails. This can be set in motion by a man pushing steadily, i.e., applying a uniform force. The speed will gradually increase from zero, and after, say, two minutes will be equal to one mile per hour. Further pushing would cause further increase of speed, or if left to itself the wagon would continue to travel at constant speed. (In an actual case, friction would soon bring it to rest, i.e., forces would come into play tending to stop the motion.) Now take again the same case, but suppose the wagon to contain a quantity of luggage. Starting from rest, the man, applying exactly the same force as before, finds that at the end of the two minutes he has obtained a speed of only half a mile per

* Sir Isaac Newton, "The greatest of Natural Philosophers" (1642–1727), was Professor of Mathematics at Cambridge and President of the Royal Society. He discovered among many other things the Law of Universal Gravitation. His great work, shortly known as the "Principia," was published in 1687.

hour. Thus the same force has been applied for the same time in each case, but in the second case the moving body has attained only half the speed. We say, then, that the inertia of the second body (wagon + load) is double that of the first (wagon alone). If another load exactly similar to the first were added, the speed attained would be only one-third of that reached in the same time in the first case, the inertia being three times that of the unloaded wagon. So that inertia depends on quantity of matter. If we take inertia to be proportional to the quantity of matter—or to Mass, as it is called—with which it is associated, we see that the wagon and each portion of luggage are equal to one another in mass. We may sum up by saying that, if a given force acts for a given time on different masses, the resulting velocities will be inversely proportional to the masses. Three times the mass, one-third the velocity. In each case the product mass × velocity is the same. This product is called Momentum. Thus, in a certain time a force gives rise to a definite momentum, whatever the mass on which it acts. Again, if the force acts for different lengths of time on any mass, the velocities and also the momenta attained—from zero in each case—will be found to be proportional to the times of action.

Mass

A force which in a certain time gives rise to twice the momentum due to another force, acting on the same mass for the same time, is said to be twice as great as the second force. Thus momentum generated is proportional to the magnitude of the force and also to the time of its application, but, these being given, it is independent of the mass of the moving body. That is:

Force applied × time of application ∝ momentum generated,

or
$$\text{Force applied} \propto \frac{\text{momentum generated}}{\text{time of application}}.$$

This last term may be called rate of change of momentum.

Law of Force

If we add to this a statement respecting the direction of the change of momentum, we have Newton's second law of motion, of which the following is a modernized version: Rate of change of momentum is proportional to force, and the change takes place in the direction in which the force acts. We have obtained the law partly by quoting results of experiments on the wagon, which could have been made more accurately with properly constructed apparatus, and partly by defining our terms, Force and Mass, to suit. Of course, having once made these definitions, we must always keep to them. The forces with which we have been dealing are really "resultant forces." For instance, if a second man applied a contrary but smaller force to the other end of the wagon, the resultant force causing motion would be the difference between the two. If the two opposing forces were equal, their resultant would be zero and no motion would take place. If frictional resistances oppose the applied force, the surplus force only, after these have been overcome, will be effective in producing motion. Newton's third law states that "To every action there is always an equal and contrary reaction; or, the mutual actions of any two bodies are always equal and oppositely directed." The man experiences the reaction of the wagon as he pushes it. If the brake were on, so that he could not move it, he would still feel the same reaction. It would be equal to the force he applied in either case.

Law of Reaction

From the second law we may obtain an expression which is often of great use. We may write

$$\text{Applied force} \propto \frac{\text{mass} \times \text{change of velocity}}{\text{time of action}}.$$

Now the change of velocity divided by the time during which that change is taking place, or the rate of change of velocity, is called the acceleration of the moving body. So,

Acceleration

$$\text{Applied force} \propto \text{mass} \times \text{acceleration},$$

or

$$F = ma.$$

In using a formula like this, we must always remember what the various symbols represent. For instance, F here means the force appled to the mass m, and a signifies the acceleration which will be produced, or the rate of change of the velocity of the body whose mass is m.* If we wish to speak only of the numerical values of the force, mass, etc., we may use the symbols **F**, **m**, **a**, and we may write—

$$\mathbf{F} = \mathbf{ma}.$$

Take a case in which a force acts for one second, and the body, originally at rest, is at the end of the second moving with a velocity of 1 cm. per sec. The acceleration (or change of velocity per sec.) is 1 cm. per sec. per sec. Its numerical

* The symbols here used represent physical quantities. The more advanced reader may refer to Appendix A on this subject.

value is 1, i.e., $a = 1$ in this special case. Now let us suppose the mass of the body to be unity, so that $m = 1$. We may take any arbitrary unit for our mass. The most convenient is the gramme. It is practically the quantity of matter in 1 c.c. of water, or in the piece of brass stamped 1 grm. in a box of weights. Supposing then this to be the mass in our example, we shall have in this case

Unit of Force

$$F = 1 \times 1 = 1.$$

But this means that the numerical value of the force in question is unity; or that the force which acting on unit mass produces unit acceleration is unit force. This unit of force is called the Dyne. If we had used the pound for our unit mass instead of the gramme, a larger unit of force would have been necessary to make the equation fit. Its name is the Poundal.

Physical Dimensions

Before we leave this equation $F = ma$ it will be convenient to call attention to the principle of physical dimensions. It is a special application of a principle which applies to all consistent thought. We apply it, without realizing it, perhaps without even knowing that such a principle exists, in everyday conversation. On being asked the distance to a certain place, one might give a true answer, say 100 miles, or an erroneous one, say 120 miles, but one would hardly be likely to fall into the error of answering 100 square miles. Such an answer, though correct so far as the mere number 100 goes, is really meaningless. The other answer, 120 miles, though inaccurate, at least is not nonsense. Now it is just because errors such as the square mile one above are as a rule so obviously nonsensical that we do not need to know any special principle of thought as a safeguard against them. But in Physics equally meaningless statements, especially in the form of equations, might very easily be made without the error being in the least obvious. The statement that the distance from X to Y is 100 square miles is a breach of a definite principle. Let us put the statement into the form of an equation:

$$\text{Distance } \overline{XY} = 100 \text{ sq. miles},$$

or

$$\text{Length } \overline{XY} = \text{area } 100.$$

Thus we are really equating a length to an area. Now our principle is that all the terms of an equation must represent quantities of the same kind. Let us take our equation

$$F = ma.$$

Now m represents mass, and a acceleration. The latter equals [change of velocity] ÷ [time], the square brackets being used to call attention to the fact that we are thinking of the *kind* of quantity only, and not its numerical value. But a change of velocity is of the same kind as the velocity itself; therefore acceleration equals [velocity] ÷ [time], or, as we may say, it has the physical dimensions of a velocity divided by a time. But velocity itself equals [length] ÷ [time]. A velocity, for instance, will be so many miles per hour, or cms. per second. It equals [distance] ÷ [time taken to travel that distance]. Thus acceleration is seen to be of dimensions $\frac{[\text{length}]}{[\text{time}]\,[\text{time}]}$; it is measured in cms. per sec. per sec. Now the right hand side of our equation has, therefore, the dimensions $\frac{[\text{mass}]\,[\text{length}]}{[\text{time}]\,[\text{time}]}$ or, as we may write, $\frac{[\text{mass}]\,[\text{length}]}{[\text{time}]^2}$; therefore by our principle we must assume that force is also of these dimensions. Then, whenever we come across force in a physical equation in future we must ascribe to it these dimensions.

Now to apply our method. A student submits an answer to a physical problem. His answer is given:

$$Ft = mv^2,$$

where t = time, v = velocity. The question is not whether this result is a correct solution of the problem submitted, but whether it makes sense. Can a quantity which has dimensions [force] [time] be equal to one whose dimensions are those of [mass] [velocity]^2? Test the dimensions of the two terms of the equation. The first term, Ft, has evidently dimensions of $\frac{[\text{mass}]\,[\text{length}]}{[\text{time}]^2} \times [\text{time}]$ or $\frac{[\text{mass}]\,[\text{length}]}{[\text{time}]}$. The dimensions of the second are $[\text{mass}] \times \frac{[\text{length}]^2}{[\text{time}]^2}$ or $\frac{[\text{mass}]\,[\text{length}]^2}{[\text{time}]^2}$. Thus the terms have different dimensions and the equation is as nonsensical as the statement that distance XY = 100 sq. miles. Had the result been $Ft = mv$, the dimensions of the two terms would have been equal, and therefore the equation would have "made sense," though of course that does not necessarily mean that it would have been the correct solution of the problem. The other was really as stupid as this:

6 horses = 3 cows × 2 pigs

which is numerically correct, but physically meaningless.

From simple principles such as those which have occupied our attention in this chapter springs the science of mechanics.

CHAPTER II

Gravitation—Earth and Moon—Weight—The Moon's Orbit—The Gravitation Constant—Mass and Weight.

WE must now pass on to consider another property which is common to all substances known to us—that of Gravitation. Newton stated that "Every particle of matter in the Universe attracts every other particle," and went on to give the law of this attraction.

Gravitation

How the particles attract we do not know. Whatever the explanation may be, each body subject to gravitational attraction behaves as though a force were applied to it, urging it in the direction of the attraction. In fact, we speak of attraction as a force, which we can treat in exactly the same way as any other force. But we are not warranted in applying Newton's law to the smallest known particles of matter. Our ideas of such particles are quite different from the ideas of Newton's time. We do not know exactly how they behave towards each other individually. Certainly large masses, such as the earth and the moon, behave as though every part of the one attracts every part of the other, with the force given by Newton's law. The law states that the mutual attraction between two particles is proportional to the product of their masses, and inversely proportional to the square of the distance between them. Thus, if the masses are m_1 and m_2 and the distance between them d, the force on either will be proportional to $\frac{m_1 m_2}{d^2}$. To calculate directly from this law in the case of the earth and the moon would be a complicated matter, for we should have to take a particle of the earth and calculate its attraction on every particle of the moon, then take a second particle of earth and repeat the process, and so on. But, fortunately, mathematics has shown that, according to Newton's law, if the two attracting bodies are homogeneous spheres, or if they are spheres which consist of homogeneous concentric layers or shells, then the total attraction between the spheres is proportional to the product of their masses divided by the square of the distance between their centres. In fact, the spheres attract each other as though their masses were entirely concentrated at their centres. Thus, if M_1 is the mass of the earth and M_2 that of the moon,

Earth and Moon

and r the distance between their centres, the mutual attraction is proportional to $\frac{M_1 M_2}{r^2}$. This force would draw the two together but for the fact that the moon is already in rapid motion in another direction.

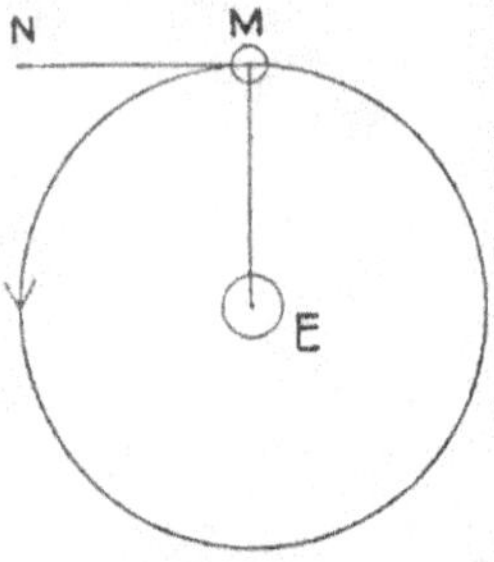

Fig. 1.

In Fig. 1 the moon M is at a given instant moving in the direction MN, with velocity v. If this motion were stopped, M would fall to earth E. If, instead of this, E's attraction were suddenly withdrawn, M by virtue of its inertia would continue to move in a straight line—in direction MN with uniform velocity v.

Again, a large mass like the earth and a small one, even the smallest with which we can practically experiment, behave as though they attracted one another according to Newton's law, though, to

Weight

be sure, we can detect the effect only on the small one—that effect is the Weight of the body. Since the earth acts as though its mass M_1 were concentrated at its centre, the attraction on the small mass, m say, near the earth's surface will be proportional to $\frac{M_1 m}{R^2}$, where R is the radius of the earth, or will be equal to $G\frac{M_1 m}{R^2}$, where G is a constant. When the value of this constant is once determined it should apply to all cases, whether the attracting masses are worlds or specks of dust—if Newton's law be true. The force or attraction $G\frac{M_1 m}{R^2}$ is, then, the weight of the body. This would diminish if the body were removed further from the earth.

The attraction of the earth on the moon, if the law holds, is $G\frac{M_1 M_2}{R^2}$, the only difference being that the mass of the moon and its distance from the earth's centre are different from the corresponding values for the small body. Now it can be verified that the same law does hold for these two cases. Let E in Fig. 2 represent the earth, and let a small body, say a stone, be thrown horizontally from a tower, P. It will describe a curve PA, being pulled down by its weight, i.e., by the attraction of the earth, and will come to rest at A. Next, let it be projected again from P with a greater velocity than before. This time it will follow the path PB. We assume no air resistance. The curves PA, PB are portions of ellipses, and the weight and the velocity of projection being known, they can be calculated. Also the time taken by the flights can be calculated. If the velocity of projection were sufficient, the body, although constantly attracted to the earth, would describe the circle PQP, and the time of flight could be calculated. If not interfered with the body would circulate perpetually around the earth with uniform speed. Now suppose the height of the tower to be increased. Assuming Newton's law to hold, the weight of the body would be less by a calculable amount, and the circular orbit would be greater. A different velocity would be required, and again the time of revolution could be calculated. If the tower were so tall as to reach to P^1, a point on the orbit of the moon, the stone, whose weight could again be calculated, would, if projected with proper velocity, describe a circle round the earth in a period which could be calculated. Now this period, calculated by means of the weight of the stone as deduced from Newton's law, is found to coincide with the actual period of the moon's revolution round the earth, thus showing that the attraction which applies to a stone at the earth's surface, giving it its weight, also applies to the moon and keeps the latter in its orbit.

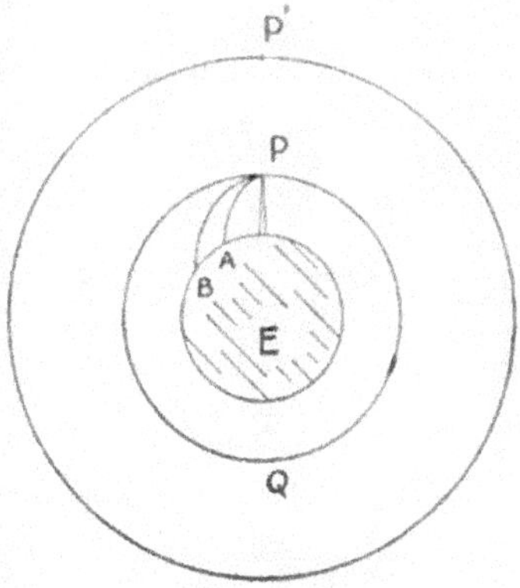

FIG. 2.

The Moon's Orbit

A question may arise here as to whether the greater mass of the moon should not cause its time of flight to be different from that of the stone in the case given above. But if one stone takes a given time to complete the circuit, two stones projected together would take the same time, and this would not be altered if the two were joined together. So three, four, or any number might be joined to form a great mass, and the time would be unaltered.

Gravitation can also be shown to exist between two small masses such as can be experimented with in the laboratory. Many experiments have been made to measure this attraction. Assuming Newton's law to hold for this case, the value of G can be determined, for here both masses are known, also the distance between them and the actual attraction as well. In a set of experiments by Poynting,* a spherical mass of about 20 kilograms was attached to a sensitive balance, and another mass of about 150 kilos brought underneath it, and the attraction, which was exceedingly small, estimated by the resulting deflection of the balance. The numerical value of G was found to be about $\frac{6.7}{10^8}$.

The Gravitation Constant

We have seen that the weight of a body of mass m_1 near the earth's surface is $G\frac{M_1 m_1}{R^2}$. This force measured in dynes, divided by the mass of the body, gives the acceleration of the body towards the earth, i.e., acceleration $= G\frac{M_1}{R^2}$. If the body falls freely from rest under the action of gravity, its velocity at the end of 1 second will be numerically equal to the acceleration. This velocity might be found directly by experiment, though other and more accurate methods can be employed. Its value near the earth's surface is about 981 cms. per sec., so that the acceleration of the body is 981 cms. per sec. per sec.

Thus, taking numerical values only :†

$$\mathbf{G}\frac{\mathbf{M}_1}{\mathbf{R}^2} = 981,$$

or

$$\frac{6{\cdot}7}{10^8} \cdot \frac{\mathbf{M}_1}{(6{\cdot}37 \times 10^8)^2} = 981,$$

for the radius R of the earth is about $6{\cdot}37 \times 10^8$ cms. This

* Dr. J. H. Poynting, late Professor of Physics in the University of Birmingham.

† Remember $\mathbf{M}_1$ is the numerical value of the mass M_1, thus we may say $M_1 = \mathbf{M}_1$ grms., and so on.

gives for the mass M_1 of the earth about 6×10^{27} grms. Also, since the weight of the body whose mass is m_1 is $G\frac{M_1}{R^2}m_1$, and since $G\frac{M_1}{R^2} = 981$ cms. per sec. per sec., we see that the weight of a body in dynes is numerically equal to 981 times its mass in grms., and, in particular, that a mass of 1 grm. weighs 981 dynes. The weight diminishes as the mass is removed further from the earth's surface.

Although the law of gravitation is found to have such wide applications, it has not been proved to apply to the smallest particles of matter, and we must be careful not to assume, on insufficient evidence, that it does so.

Mass and Weight

Before leaving the subject of gravitation it may be desirable to consider a point about which confusion often exists. We have spoken of mass as quantity of matter, and weight as a force of attraction between a body and the earth. The first is measured in grammes and the second in dynes. The first has the dimensions of [mass], and the second of $\frac{[\text{mass}]\,[\text{length}]}{[\text{time}]^2}$. The first is constant in magnitude wherever the body may be, the second diminishes as the body is removed farther from the earth's surface, and in outer space has practically no value at all. And yet these two entirely different things are very often confused with one another. The student is not altogether to blame for this. In common language the gramme and the pound are spoken of as weights instead of masses, and even in the Physics classroom the operation of determining the mass of a body is spoken of as weighing; and a box of standard masses is called a box of weights. And so far as bodies at or near the earth's surface are concerned, the weight is proportional to the mass, for we have seen that, numerically, the weight = 981 × the mass for all bodies, and this accounts to some extent for the confusion. An illustration may serve to bring home the difference between inertia and weight. Inertia, as we have seen, is bound up with the mass of a body; it remains constant however the weight may vary with changes of distance from the earth. A man striking horizontal blows with a hammer is making use of the inertia and not of the weight of the hammer. A blow struck downward is to some extent helped by the weight and one struck upward is hindered by it. If a cavity could be made in the centre of the earth, and the man and his hammer conveyed to it without injury, they would have no weight, for, although every portion of the earth would still attract them, these attractions would be distributed equally in all directions, and there would be no resultant attraction or weight. But the mass and the inertia of the hammer would remain, and blows equal to the horizontal ones referred to above could be struck.

CHAPTER III

Solids and Fluids—Elasticity—Kinetic Theory of Matter—Surface Action—Vapour—Surface Tension—Vapour Pressure—Criticism of Method—Calculation of the Pressure.

SO far we have dealt with properties common to all forms of matter. But the most casual observation will show us that different substances possess very different properties. One of the most obvious distinctions is that between solids and fluids, the latter comprising liquids, vapours, and gases. The scientific distinction between solids and fluids does not exactly coincide with the everyday one. A lump of pitch is strictly a fluid, though it is a hard and brittle substance. If it be put into a funnel, and allowed to remain long enough, it will flow like a thick, viscous liquid. A coin placed on the surface of a vessel of pitch will sink in time through the mass to the bottom. The pitch, in fact, will yield continuously to the slightest force tending to change its shape, though it may do so only very slowly. This constitutes its right to be regarded as a fluid. Every solid body will yield somewhat, even to the slightest force, but not continuously. It will yield to such and such an extent, but, however long the application of force be continued, no further yielding will take place. When the force is withdrawn, the solid will regain its original shape. Great force, however, will cause permanent distortion or fracture. The yielding which takes place in a solid, due to the application of a force, is proportional to the force, so long as this is not too great. This is a general law of Elasticity, called Hooke's law.* Both solids and fluids yield more or less to forces which tend to compress them into smaller volume. The extent of this yielding is also proportional to the applied pressure. It is not continuous, but definite in amount, and on removal of the pressure the substance regains its original volume.

Solids and Fluids

Elasticity

* Robert Hooke (1635–1703) was an English experimental philosopher and inventor.

Kinetic Theory of Matter

All matter, solid or fluid, is made up of small particles, the constitution of which we shall consider later, and these are in a continual state of agitation. The agitation of the particles of a liquid is rendered evident by what is known as the Brownian movement. In 1827 Brown, an English botanist, described a microscopical observation on minute bodies suspended in liquid. These he found to be moving continually in rapid irregular jerks, the motion being more vigorous the smaller the bodies. The reason for this movement was not understood at the time, but it is now known to be caused by irregular bombardments of the bodies by the particles of liquid. The experiment forms a very interesting confirmation of the kinetic theory of matter.

Surface Action

In solids the separate particles—or, at least, those which form the general framework or skeleton of the body—never move any considerable distance from their mean positions. In many cases of crystals the actual spacing and arrangement of these have been determined by a remarkable application of X-rays. But in liquids and gases the particles are free to roam about and visit all parts of the substance. In both solids and liquids the particles, being close together, exert strong forces on each other, but in the case of liquids these do not prevent the particles from gliding about amongst one another. They do, however, exert a strong influence in preventing a particle from leaving the liquid altogether, say at a surface exposed to the air. For, although a particle in the interior of the liquid, which is acted on in all directions by the adjacent particles, is, on the whole, free to move in any direction so far as these forces are concerned, yet a particle near the surface of the liquid—with but few others between it and the surface—is acted on by a strong resultant force towards the interior of the liquid mass. Thus it has difficulty in escaping, and will not do so unless its velocity towards the surface is great. The forces holding the particle back must not, however, be supposed to be gravitational. As we have already said, we do not know exactly the nature of these forces which act between individual particles. Those with which we are now dealing at any rate have no appreciable effect except at very minute distances. But,.be this as it may, a particle roaming about through a liquid finds it very difficult to pass across the surface layer. As it nears the surface its velocity towards the surface will be rapidly reduced, and the contrary force will increase. Only the swifter particles will escape. These will travel in straight lines till they collide with other particles or obstacles, when they will rebound and travel again in straight lines, generally with altered velocities.

Imagine a case in which a closed vessel, A, Fig. 3, contains originally nothing but a quantity of liquid, L, the space S above this being a perfect vacuum. From time to time particles will emerge from the liquid surface, and will "cushion" and "cannon" and occasionally return to "baulk"—i.e., will in their erratic flight penetrate the liquid surface again, to be held back for a while from further participation in the game. At first, the number leaving the surface per second will be greater than the number returning. But, as the population in the space S becomes greater, the number striking the surface per second will increase, and at last will equal the number leaving it. Then the population will remain stationary. If it increased, the number striking the surface would be in excess of the number leaving it in a given time, and this would cause the population to be reduced again. So a "steady state" will be reached in which a nearly constant number of particles, though the individuals will be continually changing, will exist in S. They will constitute the vapour of the liquid. The distinction between the liquid and the vapour in the vessel is due to the action of the surface layer in holding back all but a small proportion of the liquid particles. If, however, the vapour were drawn off as fast as it formed, so that no particles returned to the liquid, the whole of the liquid would in time evaporate.

Vapour

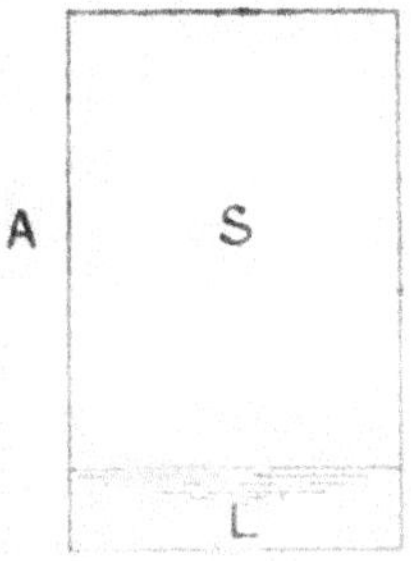

Fig. 3.

But the surface effect is not only such as to give rise to a distinct demarcation between the liquid and its vapour. To understand this, imagine a spherical drop of water, A, Fig. 4, containing, say, 1 cubic cm., the shaded portion, much exaggerated, representing the surface layer into which particles find it difficult to enter. Now suppose the drop drawn out into the form B. Evidently much more of the 1 c.c. is now contained in the surface layer than before, and therefore many particles have been forced into this layer which were originally in the

Surface Tension

body of the drop. Such a process would be resisted by the forces we have been considering, and if the drop were now

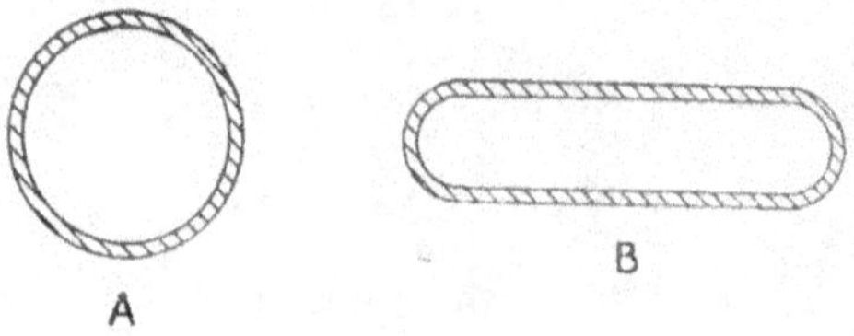

FIG. 4.

left to itself, it would tend to resume its original spherical shape—the shape in which the surface is a minimum. A liquid thus resists any attempt to increase the area of its surface; it tends to reduce this area as much as possible. In fact, the surface layer behaves very much like a film of stretched rubber. This property is known as Surface Tension. Many striking experiments can be made with liquid films in a state of tension; the commonest is the blowing of an ordinary soap bubble.

Vapour Pressure

We will now return to the vapour in the space S, Fig. 3. This vapour exerts a pressure on the sides of the vessel. To account for this we must refer again to the laws of motion. We saw that the force applied to a body is proportional to the rate of change of momentum generated, i.e., to the momentum generated per second. Using proper units, we may write: Applied force = momentum generated in 1 second. Again, if a body possesses momentum, and a force be applied so as to bring the body to rest, the same relation applies, the force is equal to the reduction of momentum brought about in one second. Thus, if a body having a momentum of 100 units (and as momentum is mass × velocity, or $\frac{[\text{mass}]\,[\text{length}]}{[\text{time}]}$, we must call the units grm. cms. per sec.) be brought to rest in 5 secs. by a force uniformly applied, this force must be equal to $\frac{100}{5}$ or 20 dynes, for the momentum is reduced by 20 units per second. As action and reaction are equal, we might as well say that the body exerts this force against that which is resisting its motion. The body in 5 seconds gives up

100 grm. cms. per sec. of momentum to the resisting body, and exerts in so doing a uniform force of 20 dynes upon it. Thus the force exerted is equal to the momentum given up per second. Now, if a stream of matter strikes an obstacle, and is thereby brought to rest, as in the case of water striking a surface AB, Fig. 5 (and falling away by gravity—this does not affect our calculation), then the force exerted on the obstacle by the stream is equal to the momentum given up by the stream (or destroyed in the stream by the obstacle) per second, the force, of course, continuing uniformly as long as the stream acts. To calculate this force it is only necessary to multiply the mass of water in grms. coming up in 1 second by the velocity of the stream in cms. per second, and the resulting force will be given in dynes.

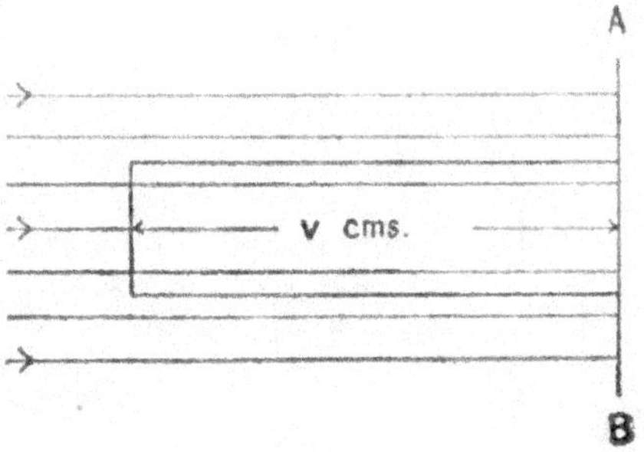

Fig. 5.

We will consider the force on 1 sq. cm. of the plate, and disregard that part of the stream falling outside this area. The force per sq. cm. is called the Pressure. We wish then to calculate the pressure on the plate. Let v = velocity of the stream in cms. per sec. Consider an imaginary cylinder of water of length v measured in cms., and 1 sq. cm. in cross sectional area. The length of the cylinder is thus taken to be numerically equal to the velocity of the stream. It cannot be physically equal to it, for the dimensions of a length are not the same as those of a velocity. We have to calculate the mass of the water coming up in 1 sec. Evidently, all the water which is in the cylinder at the beginning of any second will strike the plate during the course of that second. None outside will do so. We need to know the mass of this quantity of water. Its volume is $v \times 1^2$, measured in c.cs., and if its density, i.e., the mass per c.c., be denoted by d, the total mass will be $dv1^2$, or dv. This mass multiplied by the velocity, v, will give us the pressure required, viz., dv^2, measured in dynes per sq. cm.

Before proceeding, let us test our result by the method of dimensions. The test will not of course guarantee the accuracy of the result. The pressure is a force divided by an area (force per sq. cm.), i.e., its dimensions are those of

Criticism of Method

$$\frac{[\text{force}]}{[\text{length}]^2} \text{ or } \frac{[\text{mass}]\ [\text{length}]}{[\text{time}]^2\ [\text{length}]^2}, \text{ i.e., } \frac{[\text{mass}]}{[\text{time}]^2\ [\text{length}]}.$$

Now the product dv^2 has dimensions [density] × [velocity]^2, i.e. :

$$\frac{[\text{mass}]}{[\text{volume}]} \times \frac{[\text{length}]^2}{[\text{time}]} \text{ or } \frac{[\text{mass}]}{[\text{length}]^3} \times \frac{[\text{length}]^2}{[\text{time}]^2},$$

$$\text{i.e., } \frac{[\text{mass}]}{[\text{length}]\ [\text{time}]^2}$$

as required. But during the calculation we arrived at the expression dv as representing a mass. Now, can dv represent mass ? We were using v to represent a length in cms. and d was $\frac{[\text{mass}]}{[\text{length}]^3}$; thus dv had dimensions $\frac{[\text{mass}]}{[\text{length}]^3} \times [\text{length}]$, or $\frac{[\text{mass}]}{[\text{length}]^2}$, which certainly cannot be mass. The fact is that we have dropped an area, [length]^2, for we took the area of our stream to be 1 sq. cm. or 1^2, and then, because the factor 1^2 makes no difference to a numerical value, we left it out. But in so doing we spoilt our physical dimensions. Again, we used v to represent length in one part of the calculation and velocity in another. The result turned out to be correct in the end, and the procedure we adopted is a common one, which is often convenient and fairly safe.

Let us, however, go over our problem again on more accurate lines, taking care not to drop out any dimensions. Instead of calculating force on 1 sq. cm., let us calculate it on **A** sq. cms., where A of course has dimensions [length]^2, and consider the action of the stream for **t** secs. instead of 1. Now, taking A as area of the imaginary cylinder, and L as its length (i.e., the length which comes up to the plate in **t** secs.), the mass coming up in the **t** secs. will be dAL. $\left\{\frac{[\text{mass}]}{[\text{length}]^3} \times [\text{length}]^2 \times [\text{length}] = [\text{mass}]\right\}$. The velocity is **v** cms. per sec., and therefore the momentum given up in the t secs. is $dALv$, $\left\{\frac{[\text{mass}]}{[\text{length}]^3} \times [\text{length}]^2 \times [\text{length}] \times \frac{[\text{length}]}{[\text{time}]} = \frac{[\text{mass}]\ [\text{length}]}{[\text{time}]}, \text{ or } [\text{mass}] \times [\text{velocity}]\right\}$, which is correct for momentum, and the momentum given up per sec. is $\frac{dALv}{t}$ $\left\{\frac{[\text{mass}]\,[\text{length}]}{[\text{time}]^2}\right\}$, which is of correct dimensions for a force. But $\frac{L}{t}$ is equal to the velocity v of the stream, and it is of correct dimensions for a velocity; thus the force exerted is dAv^2. And the force per sq. cm. (or pressure) is $\frac{dAv^2}{A}$ or dv^2, which as before is of correct dimensions for pressure. We have thus found the pressure exerted by a stream of matter of given density impinging on a plate with given velocity. If a similar stream were projected from the plate, say by a series of fine jets, there would be a reaction on the plate just equal to the force calculated above. So that, if the stream impinging on the plate rebounded with the same velocity, the total pressure on the plate would be double that which we have calculated.

Calculation of the Pressure

Now to apply all this to the problem of vapour pressure. The particles move in all directions and with various velocities; but if we wish to calculate the pressure which, due to impacts and rebounds, they exert on the walls of the vessel, we must simplify our problem by supposing, first, that they all move with a mean velocity v (say **v** cms. per sec.), and then that they move in paths either parallel or perpendicular to the wall on which the pressure is to be determined. Thus we may suppose that $\frac{1}{6}$ of the particles move in a stream with velocity v in direction A, perpendicular to the wall XY, Fig. 6; another $\frac{1}{6}$ in direction B; a third and fourth in directions C and D; and the two remaining portions in directions towards and away from the reader, perpendicular to the plane of the diagram, all with velocity v. The effect will be nearly the same, so far as the plate XY is concerned, as in the more complicated actual case. Each of the six streams is

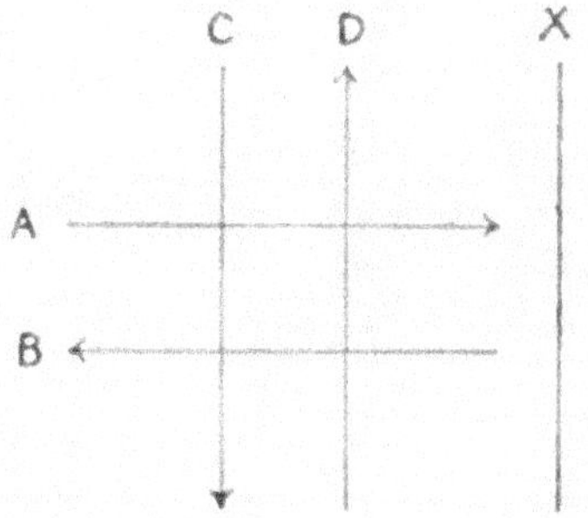

FIG. 6.

supposed to fill the whole of the vessel and to move through the others—no collisions taking place. Evidently, only the first and second streams will produce an effect on XY, and these will produce equal pressures on it. We need, then, only calculate the pressure due to the first stream, and double this to get the whole pressure on XY due to the vapour. Just as in the case of the stream of water we find the pressure by multiplying the density by the square of the velocity. But if the density of the vapour in grms. per c.c. is ρ, the density of one stream is only $\frac{1}{6}\rho$, and therefore the pressure due to this is $\frac{1}{6}\rho v^2$. Doubling this, we find that the "vapour pressure" is equal to $\frac{1}{3}\rho v^2$. This pressure is exerted on every surface with which the vapour is in contact, including the surface of the liquid. Indeed, we already know that as many particles leave the liquid surface in a given time as re-enter it, and the effect of a particle striking the surface and sticking, and of a new particle leaving the surface, is just the same, so far as pressure is concerned, as that of a single particle striking the surface and rebounding.

CHAPTER IV

Kinetic Energy—Potential Energy—Expression for Kinetic Energy—Heat—Critical State—Temperature—Boyle's Law—Gas Law—Gas Thermometer—Latent Heat—Joule's Equivalent—Specific Heat.

IT is now necessary to introduce a new conception, that of Energy. A moving body possesses, by virtue of its motion, Kinetic Energy, the quantity of which depends partly on the mass of the body and partly on its velocity. Each particle of vapour possesses its own kinetic energy, which alters in amount every time a collision causes a change in its speed. But energy can exist in other forms, and can be transferred from body to body, just as one moving particle may collide with and set into motion another, so giving up part or all of its kinetic energy.

Kinetic Energy

If a heavy body is to be lifted to a certain height, work must be done, or energy expended, equal to the weight of the body multiplied by the height through which it is raised. The upward force to be applied must be equal to the weight of the body or, what is the same thing, to the force of gravity acting upon it. If the distance traversed is small, we may consider this force to be uniform. Let the weight, or the lifting force, be **w** dynes, and the height raised **h** cms.; then the energy expended is equal to **wh** units. These units are called Ergs. The energy, however, is not lost; it is stored in the body (strictly in the gravitating system—earth and body). It is said to exist in the form of Potential Energy.

Potential Energy

Now suppose the body to be allowed to fall freely under the action of gravity through **h** cms. Gravity pulls, and the velocity of the body continually increases, exactly as was the case with the wagon of Chapter I, to which, also, a uniform force was applied. At the end of the h cms. the body moves with its maximum velocity v, say, just before striking the ground. What has become of the potential energy? It exists as kinetic energy in the moving body. This kinetic energy could, by a suitable mechanism, be transferred to another body of equal weight, and made to lift it exactly **h** cms.—at least if mechanisms could be made to work perfectly and without friction. Thus the kinetic energy is equal in amount to the original potential energy. Let us obtain an expression for the kinetic energy. First we must determine the velocity of the falling body at a depth of h cms. from its highest point. During the fall, a force of **w** dynes has been acting, and we know, therefore, that the acceleration, say g, of the body will be given by the equation $w = mg$.* Since weight and mass are proportional for all bodies at the earth's surface, g is the same for all bodies at the surface—subject to a small correction due to the fact that the earth is not exactly spherical. It is less for bodies further removed, for, while m remains constant, w diminishes. Now, as we saw in Chapter II, since the body starts from rest, it is clear (the acceleration being the rate of growth of velocity, or the growth of velocity per second) that at the end of one second the velocity in cms. per sec. will have become numerically equal to g. That is, the velocity at the end of 1 second will be numerically equal to the acceleration of the body. As the acceleration remains constant throughout the fall, being always $\frac{w}{m}$, it follows that at the end of the 2nd second the velocity will have become numerically equal to $2g$, and so on. Or in time t from the start, the velocity will be gt; thus we obtain: $v = gt$.

In the example proposed, however, we do not know the time of fall, but only the distance h. We must therefore connect h and t by an equation, and then we can determine the velocity at the end of the fall h. Now, if a body falls for time t, with uniform velocity v, the distance travelled will be vt. But if it falls from rest with uniformly increasing velocity, the final velocity being v, the distance will be (mean velocity) × (time), i.e., $\frac{1}{2}vt$. In our example, then, we have $h = \frac{1}{2}vt$, v being the final velocity of the body which started from rest, t the time of falling, and h the distance fallen. Now we are in a position to determine what we set out to find, viz., the velocity of the body at the end of its fall of h cms. For:

$$v = \frac{2h}{t} \text{ and } t = \frac{v}{g}; \therefore v = \frac{2h}{v/g}, \text{ or } v^2 = 2gh.$$

* g is the $G\frac{M}{R^2}$ of Chapter II.

This gives the square of the velocity after fall h from rest. But the energy wh or mgh, which was in the potential form, has now become transformed into the kinetic form, so we must express mgh in terms of the mass and velocity of the body. Or we must write an equation of this form: (Potential energy at start) = (Kinetic energy at finish), i.e., mgh = (an expression in terms of mass and final velocity of body).

Expression for Kinetic Energy

But since the equation must be true, and since $h = \frac{v^2}{2g}$, we can only write

$$mgh = mg\frac{v^2}{2g},$$

or

$$mgh = \tfrac{1}{2}mv^2,$$

which shows us that the kinetic energy of the moving body is equal to $\frac{1}{2}mv^2$.

Now any body of mass m moving with velocity v has this kinetic energy, whether the velocity be produced by gravity or not, and whatever the direction of motion may be. Kinetic energy has nothing special to do with gravity; it may equally well be obtained by means of any other force, such as force applied to a cricket ball by the hand or to a wagon by a man pushing it.

Every particle, then, of a vapour is endowed with kinetic energy, and the whole kinetic energy of the vapour is simply the sum of the kinetic energies of the particles. It does not necessarily follow that all the energy of the vapour is in the form of "kinetic energy of translation." The particles themselves may have energy, due to rotations and vibrations on their own account, in addition.

In order to deal practically with the energy of the vapour we cannot consider each particle separately, but must take the energy as a whole. This is called the Heat of the vapour. Besides heat, the vapour may possess other forms of energy, such as chemical energy. But unless chemical action takes place this will remain in the potential form. Liquids and solids, whose particles are in rapid motion, also possess heat energy.

Heat

Consider a vessel like that of Fig. 3, and suppose it to contain only liquid and vapour, the latter in a saturated condition, i.e., in the steady state referred to above. Now let us increase the kinetic energy of the whole: vessel, liquid, and vapour. To do this we may surround it with a large vessel containing a vapour, the particles of which have greater velocities than those of the inner vessel. Or a constant stream of such vapour may be passed through the outer vessel. Now these high speed particles continually strike the outside of A, Fig. 7, and communicate some of their kinetic energy to the particles composing it. These in turn pass on energy to the liquid and vapour within, thus speeding up their particles. This goes on until the inner ones possess on the average as much kinetic energy as those outside. We have, in fact, heated the vessel A and its contents by means of a hot vapour jacket, say a steam jacket.

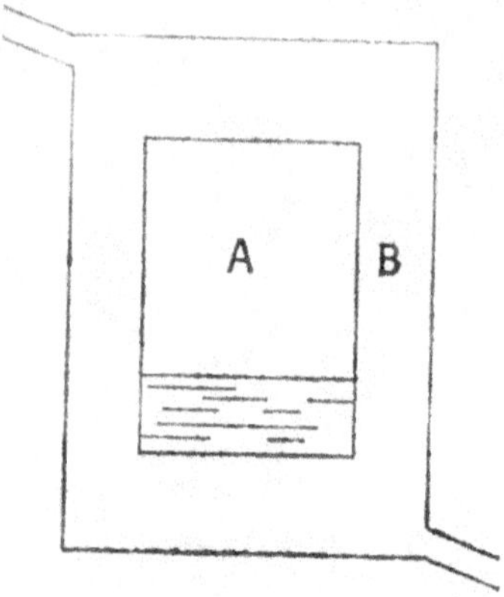

Fig. 7.

A new state of affairs now exists in A. Liquid particles in greater numbers than before are able to penetrate the surface and escape into the vapour. This increases the population in the space above the liquid, until a new steady state is reached, in which the number of particles in the vapour is greater, and consequently the quantity of liquid less than before. The particles also move with greater velocity than before, and so the streams impinging on the walls of the vessel are both denser and swifter, thus the pressure is greater; in fact, the vapour pressure has been increased. If more and more heat be applied, the vapour will become denser and denser, and the particles will more and more easily penetrate the surface layer. The surface tension will diminish, and finally all signs of a surface of separation between liquid and vapour will disappear. The "Critical State" has been reached, and the vessel is filled with a homogeneous mass of the substance. If more heat be applied, the condition of the substance is said to be that of a gas. Below the critical point, liquid and vapour can exist; above it, only gas. Below the critical point, pressure applied to the vapour will cause it to condense to the liquid form; above it, no amount of pressure will do so. Atmospheric air in its ordinary condition is well above its critical point, and cannot be liquefied by the mere application

Critical State

of pressure. To produce liquid air, much heat must be extracted and great pressure applied.

The kinetic theory, which we have applied to vapour, is also true for gases. Consider a vessel full of gas. Its particles give rise to a pressure equal to $\frac{1}{3}\rho v^2$, the symbols having the same meaning as before. In order to be able to express its state with regard to heat, we must introduce the term Temperature. Now, we possess a special sense by means of which we can perceive variations of hotness or coldness. If, for instance, we plunge a hand into the mass of gas, we can estimate directly (but roughly) its temperature or its degree of hotness. We cannot estimate the quantity of heat in the gas—that is a matter of the kinetic energy of the particles; heat and temperature must not be confused. There is vastly more heat in a cubic foot of hot water than in an equal volume of water vapour standing over it; yet the temperature of the two may be, and would be under conditions like those of Fig. 3, the same. Temperature, in fact, is analogous with Level. Just as water will flow from a tank at high level to one at lower level, if a channel be opened between them, so heat will flow from a body at high temperature to another at lower temperature, in contact with it. In fact, the use of the words high and low in connexion with temperature suggests this analogy. The heat in the case of Fig. 7 flowed from the outer high temperature jacket to the inner vessel, which was at lower temperature, till the latter was raised to the temperature of the jacket. Considering, then, our vessel full of gas, we must ascribe to it a definite temperature, say T. If the gas be heated, the kinetic energy will be increased, i.e., the value of $\frac{1}{2}mv^2$ for each particle will on the average be increased. Since $\frac{1}{2}m$ remains the same, the value of v^2 must be greater than before. At the same time T is increased, and we may agree to consider the temperature T to be proportional to the average $\frac{1}{2}mv^2$ of the particles, say, for instance

Temperature

$$T = k\tfrac{1}{2}mv^2,$$

where k is some constant whose actual value we need not specify.

If the velocity v were reduced to zero (assuming this were possible), T would be O, or the temperature of the gas would be zero. This state of things is beyond the limits of actual experiment. Of course the pressure of the gas would at the same time be reduced to zero, for there would be no impacts.

Let us suppose that a given mass M of gas is enclosed in a vessel, Fig. 8, and that the volume V can be altered by means of the air-tight piston Q. The pressure outside Q must be equal to the pressure of the gas, supposing the piston to be without weight and capable of moving without friction. Let P be the pressure of the gas, and T its temperature, which we will call absolute temperature, because its zero corresponds to the state in which no heat at all would exist. Also let the density of the gas in grms. per c.c. be ρ; then $\rho = \frac{M}{V}$, or for the given mass density is inversely proportional to volume. If the temperature T be kept constant, and the pressure be changed, say by weighting the piston, from P_1 to P_2, the volume will change, say, from V_1 to V_2. Now

Boyle's Law

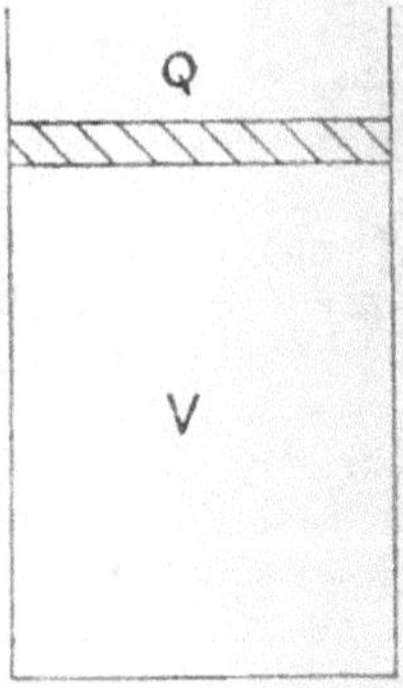

FIG. 8.

$$P_1 = \tfrac{1}{3}\rho_1 v^2 \qquad P_2 = \tfrac{1}{3}\rho_2 v^2,$$

or

$$P_1 = \tfrac{1}{3}\frac{M}{V_1}v^2 \qquad P_2 = \tfrac{1}{3}\frac{M}{V_2}v^2.$$

Thus, on dividing, we have

$$\frac{P_1}{P_2} = \frac{V_2}{V_1},$$

i.e., volumes vary inversely as pressures. Thus, if P_2 be 3 times P_1, V_2 must be $\tfrac{1}{3}$ of V_1. This is Boyle's law. Boyle* obtained the result experimentally. The law states that, if the temperature of a mass of gas be kept constant, the volume varies inversely as the pressure. The above equation can also be put into the form

$$P_1V_1 = P_2V_2,$$

that is, the product pressure $\times$ volume is the same for both conditions of the gas. It will be the same for other conditions (at constant temperature), so we may write generally

$$PV = \textit{constant}$$

for a given mass of gas at constant temperature.

When a gas is compressed at constant temperature, no change takes place in the energy of the particles, the increase of pressure in the gas being due to the fact that the impinging

* Robert Boyle, 1627–1691. A celebrated natural philosopher. Born in Ireland.

streams are denser than before (the volume of the gas being less), and not to any increase in the velocity of these streams. It is true that during compression work is done on the gas as the piston is forced down ; this would heat the gas and so raise its temperature. But we have assumed "constant temperature," and so we must suppose the heat to be allowed to escape before the final measurements of pressure and volume are made. Thus the energy put in has passed off in the form of heat, and still exists somewhere outside the gas.

If, next, the volume of the gas be kept constant and the temperature varied, the pressure will also vary. Let the absolute temperature be changed from T_1, to T_2 ; the pressure will change say from P_1 to P_2 (these values having nothing to do with P_1, and P_2 above). Now

$$T_1 = \tfrac{1}{2}kmv_1^2 \qquad T_2 = \tfrac{1}{2}kmv_2^2$$

and

$$P_1 = \tfrac{1}{3}\rho\, v_1^2 \qquad P_2 = \tfrac{1}{3}\rho\, v_2^2$$

(where ρ the density is the same for P_1 and P_2, since the volume is unaltered, and therefore the number of grms. per c.c. is unaltered). Thus, substituting, we find

$$T_1 = \tfrac{1}{2}km\frac{P_1}{\frac{1}{3}\rho} \qquad T_2 = \tfrac{1}{2}km\frac{P_2}{\frac{1}{3}\rho}$$

and by division

$$\frac{T_1}{T_2} = \frac{P_1}{P_2},$$

or, pressure is proportional to absolute temperature if volume is constant.

Again, if the temperature be varied and the pressure kept constant, the volume will vary. We have as before

$$T_1 = \tfrac{1}{2}kmv_1^2 \qquad T_2 = \tfrac{1}{2}kmv_2^2.$$

We may divide at once, and obtain

$$\frac{T_1}{T_2} = \frac{v_1^2}{v_2^2}.$$

Now, since pressure is the same in both cases,

$$\tfrac{1}{3}\rho_1 v_1^2 = \tfrac{1}{3}\rho_2 v_2^2$$

therefore

$$\frac{v_1^2}{v_2^2} = \frac{\rho_2}{\rho_1}$$

and

$$\frac{T_1}{T_2} = \frac{\rho^2}{\rho_1},$$

or, since volume varies inversely as density, at constant pressure

$$\frac{T_1}{T_2} = \frac{V_1}{V_2},$$

showing that volume is proportional to absolute temperature when pressure is kept constant. If now all three quantities vary together, say from $P_1V_1T_1$ to $P_2V_2T_2$, we can combine the three results and write

$$\frac{P_1V_1}{T_1} = \frac{P_2V_2}{T_2}$$

or

$$\frac{PV}{T} = \textit{constant}.$$

Gas Law

Thus we arrive at the gas law; the product of pressure and volume divided by absolute temperature for any given mass of gas remains constant, however the separate quantities P, V, and T may vary.

Gas Thermometer

If we keep constant the volume of a mass of gas, say in a glass bulb, and measure its pressures when it is immersed in a bath at different temperatures, the different pressures recorded, being proportional to the absolute temperatures of the gas, will serve as indications of the different temperatures of the bath. Suppose the bulb to be immersed first in melting ice, and then in boiling water. The pressures noted will represent the temperatures of the two baths. If now the bulb be immersed in warm water, and the pressure be found to be exactly midway between the other two, the temperature of the bath is said to be midway between the temperatures of the other two baths. Instead of taking the absolute zero for our zero of temperature, we might take the temperature of melting ice as zero, and then take that of boiling water as 100°. Intermediate pressures in our " constant volume gas thermometer " would indicate intermediate temperatures. Lower pressures than that obtained with the melting ice would indicate temperatures below zero on this (the Centigrade) scale, and zero pressure would indicate the absolute zero of temperature, which would be at −273° on the Centigrade scale. Plotted graphically, the vertical lines, Fig. 9, represent pressures given by the thermometer for 0° C. and 100° C., and the straight line AB indicates the way in which pressure changes with temperature. This line cuts the " axis of temperature "—the horizontal line in the diagram—at a point corresponding to −273° C. This is the absolute zero. The straightness of the line indicates the fact that pressure is proportional to absolute temperature. Our temperature T, then, represents Centigrade reading plus 273.

The absolute temperature of melting ice is therefore 273, and that of boiling water 373. The absolute scale is much more convenient for theoretical calculations than the Centigrade scale, for the latter has a purely arbitrary zero, which does not correspond to the state of a body with no heat. The absolute zero cannot be reached experimentally, but liquid helium has been prepared and reduced to a temperature of 3° absolute.

To connect what we have said with regard to a gas to the case of a vapour, we must remember that a vapour in contact with its liquid is in the saturated state, and that for a given temperature the vapour pressure is constant. If the volume be increased, or decreased, evaporation or condensation will take place until the original pressure is regained. So the

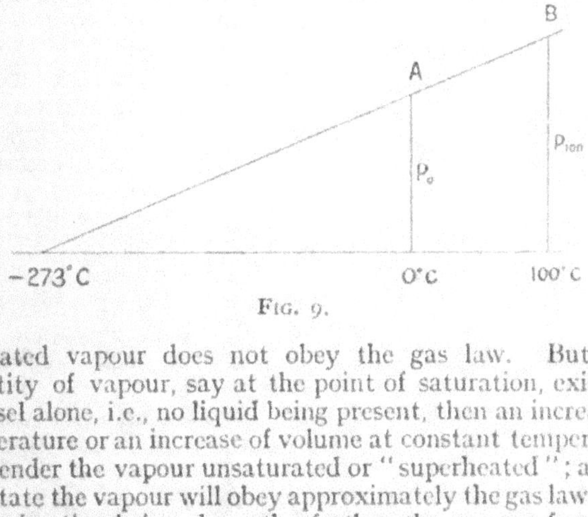

FIG. 9.

saturated vapour does not obey the gas law. But if a quantity of vapour, say at the point of saturation, exists in a vessel alone, i.e., no liquid being present, then an increase of temperature or an increase of volume at constant temperature will render the vapour unsaturated or "superheated"; and in this state the vapour will obey approximately the gas law—the approximation being closer the farther the vapour from the point of saturation. Indeed, a superheated vapour is practically a gas, but is not generally called a gas unless its temperature is above the critical temperature for the substance.

Latent Heat

When evaporation takes place from a liquid only the faster particles escape, and so the liquid is left with a greater proportion of slow ones. This means that the liquid is cooled by evaporation. If evaporation is to proceed regularly, and the liquid to maintain its temperature, heat must be continually supplied. This heat, which does not raise the temperature of the liquid to which it is given, is called Latent Heat. Heat which raises temperature is called Sensible Heat. A definite quantity of heat is necessary if a given mass of liquid is to be evaporated at a given temperature. The quantity varies from liquid to liquid, and also from temperature to temperature for the same liquid. Latent heat is also absorbed when a solid melts. The same quantity of heat is given up when a vapour condenses as was required to evaporate it at the same temperature. The same applies to the case of a freezing liquid.

We have seen that heat is a form of energy. It may be transformed into mechanical energy and vice versa. Water may be warmed by agitating it with paddles in a vessel. The mechanical energy expended in rotating the paddles against the resistance imposed by the water may be measured in ergs. The resulting heat, assuming no loss of energy to occur, will consist of that number of ergs of energy. The erg is the work unit, or mechanical unit, of heat, as it is of any other kind of energy. But this unit is not convenient practically. The practical unit of heat is the Calorie. The calorie used in Physics is the quantity of heat necessary to raise 1 grm. of water 1° C. Suppose a number of ergs of energy to be expended (or work done) in the paddle arrangement, and 1 grm. of water to be heated thereby 1° C., or, say, 10 grms. heated 1/10° C. We have expended so many ergs and have gained 1 calorie of heat. The number of ergs necessary would be about 42,000,000. Thus 42,000,000 ergs of energy are equivalent to one calorie of (heat) energy. The erg and the calorie are both energy units, but the calorie is used only in the case of heat. So 1 calorie $= 4.2 \times 10^7$ ergs. This number is called the Mechanical Equivalent of Heat, or Joule's Equivalent, the first experimental determination of it being made by Joule.* It is usually represented by the letter J. If W represents a quantity of energy in ergs, and H the corresponding quantity of heat in calories, we may write

Joule's Equivalent

$$W = JH.$$

Now, whatever form of energy is transformed into heat, the same number of ergs is required to produce one calorie, and in fact, however energy may be transformed, its quantity

* James Prescott Joule, 1818–1899. Born at Salford. A distinguished experimenter, to whom is largely due the modern theory of the Conservation of Energy.

always remains the same. This is one of the general principles of Science and is called the doctrine of the Conservation of Energy. We have no means of destroying or of creating energy. We may alter its form in many ways, but not its quantity.

The calorie depends on a particular substance, water. If we take any other substance, we can, by experiment, determine the number of calories necessary to raise 1 grm. of it 1° C. This number represents what is called the Specific Heat of the substance. Suppose it to be **S** for a certain substance. If, then, 1 grm. of the substance be raised or lowered 1° C., it gains or loses **S** calories. If **m** grms. be raised or lowered **t**° C., the number of calories gained or lost will be **mSt**. Latent heat is also expressed in calories. The heat required to evaporate 1 grm. of water at 100° C. is about 540 calories, and that required to melt 1 grm. of ice is 80 calories.

Specific Heat

CHAPTER V

Attraction—Repulsion—Electricity—The Lodestone—Magnets—A Difference—Law of Force—Choice of Units—Another Method—Conductors and Insulators—A Comparison—The Electric Current—Flow—Level, Temperature, and Potential—Law of Flow—Nature of Electric Current.

WE have spoken of the universal attraction of gravity, and have considered some of its consequences. It remains for us to investigate some other kinds of attraction which bodies exert on one another in special circumstances.

Attraction A rod of ebonite, after being rubbed with fur, is able to attract bits of paper and other light objects, as is also the fur which has been used as a rubber. The fur and the ebonite also attract one another, as can easily be demonstrated if the ebonite be suspended by a thread attached to its middle. But one rubbed rod of ebonite *repels* another rod which has been similarly treated. A glass rod rubbed with silk attracts the ebonite rubbed with fur. This effect, even if the rods be only slightly rubbed, is gigantic compared with the gravitational attraction between them. The rods and rubbers are said to be "Electrified," or charged with Electricity.

Repulsion

Electricity It is necessary thus to give names, but we must remember that to give names is not to explain phenomena. Although much is known about electricity, its ultimate nature is not understood, and probably never will be. It is clear that, using the word in the above sense, there are two kinds of electricity, for, though glass and ebonite both attract bits of paper, yet glass attracts ebonite, while ebonite repels ebonite. The ebonite and glass must therefore be charged with different kinds of electricity. Two rubbed glass rods repel each other; thus two bodies charged like glass rubbed with silk (or, say, charged positively) repel; and two bodies charged like ebonite rubbed with fur (or charged negatively) also repel. But a positively charged body and a negatively charged body attract one another. No other kinds of electricity have been discovered.

Now, although we speak of two kinds of electricity, we must remember that we are as yet only giving names. We say nothing about the nature of electricity. The one kind might be merely an absence of the other kind, so far as the foregoing experiments could show.

The Lodestone

Again, certain bodies called lodestones have been discovered which attract fragments of iron. The attracting power is not uniformly distributed over the surface of the stone, but is concentrated more particularly in certain parts. Each lodestone has at least two such parts. A particular part of a second lodestone will attract some of these and repel others. If there are only two, it will attract one and repel the other. A steel needle or rod rubbed several times, in one direction only, with a lodestone, acquires the same properties, its active parts or Poles being near the ends. The steel has been "Magnetized" and is called a Magnet. If two magnets be made in exactly the same way, the similar poles (say those at the ends where the rubbing started) will repel one another, and dissimilar poles attract.

Magnets

In that respect there is a likeness between the magnets and the electrified bodies, but, while the ebonite rubbed with fur was negatively charged only, and the fur positively charged only, every magnet possesses magnetism of both kinds—no magnet can be made which has one kind of magnetism only. If the magnet be broken in two, each half becomes a complete magnet, two new poles forming at the new ends. It may be noted that, while both attractions and repulsions occur in the case of electrified and magnetized bodies, only attraction occurs in the case of gravity. No such thing as gravitational repulsion is known.

A Difference

Now, bodies may be electrified to different amounts, and magnets made of different strengths, as the varying values of the attractions and repulsions obtained readily show. So the electric and magnetic charges should be measurable, and expressible in terms of suitably chosen units. If we denote the quantity of electricity, or the electric charge, on a small body by e and that on another by e_1, the force acting between them will be $\frac{ee_1}{r^2}$, where r is the distance between them. This law, which is similar to the law of gravitation, is accurately true only if the bodies are very small compared with the distance r, though, if we

Law of Force

consider each little bit of charge separately, we can apply the law however large the charged bodies may be. The law can be verified by experiment. If magnetic charges m, m_1 or poles of strengths m, m_1 be substituted for electric charges, the magnetic attraction or repulsion will be $\frac{mm_1}{r^2}$. In all these cases the inverse square law applies. In the electric and magnetic cases the force will be repulsion if the charges are like, and attraction if they are unlike. If we call a positive charge $+e$ or $+m$, and a negative one $-e$ or $-m$, the resulting force will be a repulsion if positive, or an attraction if negative. Thus the force between two small bodies 5 cms. apart, with charges + 10 and − 20 units of electricity respectively, or the force between two magnetic poles 5 cms. apart, of strengths + 10 and − 20 units, will be

$$\frac{(+10)(-20)}{5^2} = -8 \text{ dynes attraction.}$$

In these formulæ for electric and magnetic attraction or repulsion no constants like the G in the gravity formula are used, because the units of electricity and magnetism are chosen of such magnitude as to do away with the necessity for these. For instance, the unit of electric charge is chosen so that two such units at a distance of 1 cm. apart repel one another with a force of 1 dyne, and similarly in the case of magnetism.* (In all the above we assume the measurements of force to be made in air. If other media were interposed between the bodies, the values of the forces would, in general, be altered.)

Choice of Units

Electric attractions or repulsions are, however, not confined to bodies which have been excited by friction. If a series of glass vessels containing dilute sulphuric acid, and fitted with plates of zinc and copper joined by copper wires, be arranged as in Fig. 10,† the terminal wire A will attract the other terminal B. The effect will be very slight, but it may be rendered perceptible

Another Method

* Unfortunately, two systems of units are here involved—the Electrostatic system in the case of e and the Electromagnetic system in the case of m. The electric charge can, however, be expressed in electro-magnetic units, and pole strength in electrostatic units when it is more convenient to do so. We shall not enter into the question of the *dimensions* of electric and magnetic quantities.

† The Galvanic or Voltaic Battery due to L. Galvani, 1737–1798, Professor of Anatomy at Bologna, and A. Volta, 1745–1827, Professor f Natural Philosophy at Pavia.

by connecting the terminals to two light and delicately suspended metal plates placed near to one another. If two such batteries be arranged, the "zinc" terminal of the one may be caused to repel that of the other and likewise the two "copper" terminals. But these terminals are metallic, and rubbed metals do not show any sign of electrification. How can this be accounted for? It is true that a metal rod *held in the hand* and rubbed with fur shows no electrification. But the contrary is the case if it be held by an ebonite handle. It can then be electrified by friction. On touching the rod with the finger, or with another metal rod held in the hand, the electrification disappears. But not so if it be touched with an ebonite rod.

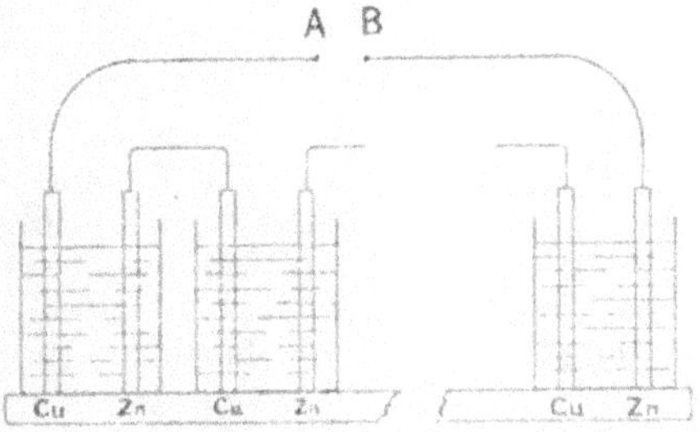

FIG. 10.

Conductors and Insulators

Imagine two metal rods to be set up on ebonite stands some distance apart, and to be connected by a wire. After one of the rods has been electrified by friction, part of the charge will be found on the other rod. If a thread of silk had been used instead of the wire, no charge would have been found on the second rod—all would have remained on the first. The wire is said to be a Conductor of electricity, the silk a Non-conductor or Insulator. The hand and body are also conductors, likewise the earth. Ebonite and glass are insulators, if thoroughly dry. A conductor, then, can be electrified by friction, but if the electricity is to be retained the conductor must be supported by an insulating stand or handle; for, if even one point of it be touched by an earth-connected conductor, such as the finger, the electricity from every part of the charged conductor will flow to that point and escape by the other conductor to earth. If one point of a charged ebonite rod be touched, the electricity from that part will escape, but that on the rest of the rod will not be able to flow to the touched point, and so will be retained on the rod. In the case of the battery, the metal terminals were insulated by means of the glass vessels, the outer surfaces of which were dry.

A Comparison

The electrification produced by the battery is not different in nature from that produced by friction. The terminal connected to the zinc is negatively, and that connected to the copper positively, electrified. An experiment could be devised, though some precautions would be necessary, in which rubbed ebonite would repel the former and attract the latter.

The Electric Current

Now let the terminals of the battery be connected by a conductor, such as a piece of copper wire. An Electric Current will flow along the wire. This current (again we are giving names which do not at all explain the nature of what really goes on) is not essentially different from that which flowed along the wire joining the metallic rods considered above. But in this case the flow is continuous. Now, flowing is characteristic of fluids. Is electricity a fluid ? According to one old theory—the two-fluid theory—positive electricity consisted of one kind of fluid and negative electricity of another. Another theory supposed that only one kind of fluid-electricity existed, the second kind of electrification being accounted for by a deficit of this fluid on the body concerned. It will be worth while to consider a little further the phenomenon of flow, which we have already come across in the case of fluids and of heat.

Flow

The three cases of flow with which we are concerned are illustrated in I, II, and III, Fig. 11. I represents flow of water, II, of heat, and III, of electricity. In I the inlet a and overflows b and c are intended to keep the levels in cisterns A and B constant. The flow with which we are concerned is that in the pipe C D. In order that the flow may take place, the levels in A and B must be different. The flow is conditioned by level-difference. The greater the level-difference the greater the flow. (The word flow or flux is often used to denote the quantity flowing per unit time.) In II we have a copper rod dipping into boiling water at one end and into a mixture of ice and water at the other. Heat flows (the process is called conduction of heat) along the rod from C to D. What really happens is that vibrations are passed on from particle to particle along the rod. The necessary condition for conduction of heat along the rod is a temperature-difference. The greater the temperature-difference the greater the flow of heat. In III electricity flows from the positive to the negative terminal of the battery along the connecting wire. In saying this we are imagining a one-fluid theory of electricity, the positive electricity representing the fluid, so we suppose positive electricity flows along the wire from copper to zinc. We shall deal later with a more modern theory, but for practical purposes we always

Level Temperature and Potential

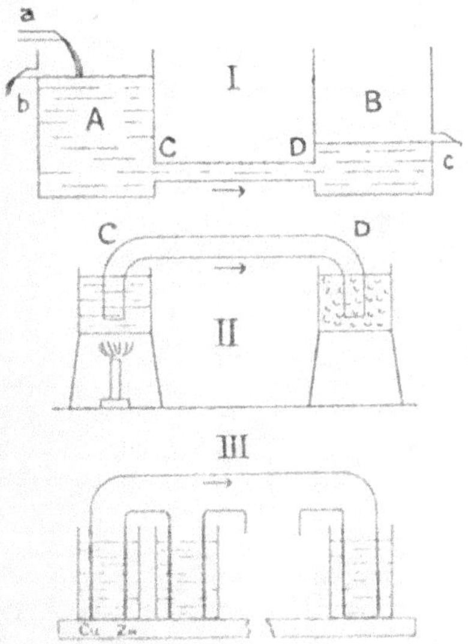

FIG. 11.

imagine electricity to flow in the direction indicated. Now, if water-flow needs water-level-difference, and heat-flow needs temperature-difference, or say, heat-level-difference (for we may look upon temperature as a kind of heat-level), what is the condition of electricity-flow ? We may say, by analogy, electricity-level-difference, or difference of electrical-level or, as it is called, difference of Electrical Potential, or more shortly, Potential-difference. Again, to give a name is not to explain, but this idea of potential is a very important one in electricity. Just as every drop of water must be at some particular level, and every little quantity of heat at some particular temperature, so every little electric charge must be at some particular potential. It always, if positive, tends to flow in such a way as to get to a lower potential. Getting to a lower potential does not involve any change in either the nature or quantity of electricity, any more than getting to a lower level involves a change in the nature or quantity of water.

In all the cases we have considered a conductor for carrying the flow is necessary. We have chosen first a pipe, second a copper rod, and third a copper wire. The flux will in each case depend, apart from level- or temperature- or potential-difference, on the efficiency of the conductor. Thus with a wide pipe in I, more water will flow than with a narrow one. And more will flow if the pipe be short than if it be long. We might speak of the Conductance of the pipe and say that a short wide pipe has greater conductance than a long narrow one. Or of the Resistance of the pipe, meaning the inverse of the above. The narrower and longer the greater the resistance. In II we might substitute other materials for copper, and for simplicity keep to the same dimensions in each case. We should find very different quantities of heat flowing in the different cases. More would flow through copper than through iron, and more through iron than through glass. The Thermal Conductivity of copper is greater than that of iron, and of iron than of glass. In III the flow of electricity depends greatly on the dimensions and material of the conductor. The flow is great along a thick and short copper rod, less along thin long copper wire, still less along a similar wire of iron, and zero along a rod of glass. The resistance of the iron wire is greater than that of the similar copper wire, and that of the short copper rod is less. The resistance of glass or of any other non-conductor may be said to be infinite.

Suppose we find that a certain flow takes place along a conductor whose ends are maintained at a certain level-temperature or potential-difference. Keeping this difference constant, we substitute another conductor and find that the flow is halved. We may say that the resistance of the second conductor is twice that of the first, or its conductance half that of the first. If the flow be reduced to one-third the resistance of the conductor will be increased to three times that of the first, or its conductance reduced to one-third. If D represents the difference,

Law of Flow

F the flow or flux, and R the resistance, then

$$F \propto \frac{D}{R}.$$

[In the case of water flow, this formula applies to long narrow pipes. The formula applies to heat conduction, but the term resistance is not generally used in this connexion, the flow being calculated in terms of the thermal conductivity of the conductor. It applies strictly to flow of electricity and embodies what is known as Ohm's law.*]

Now in the case of fluids we know that the flow is a flow of matter, while in the case of heat the flow is one of energy.

Nature of Electric Current

Is electricity matter or energy? When water flows the actual particles travel along with the flow. The flow of heat along a rod consists of a passing on of energy from particle to particle. Of which character does the flow of electricity partake? We must try to gain an idea of the modern theory and of some of the steps by which it has been reached.

* G. S. Ohm, 1787–1854. Physicist, Professor at Munich.

CHAPTER VI

Conduction through Gases—Cathode Rays—Electrons—Structure of the Atom—Ions—Millikan's Experiment—Charge on an Electron—Frictional Charges—Theory of Conduction—Heating Effect—The Resistance—The Energy—Choice of Units.

IF a glass tube A B, Fig. 12, containing a suitably rarefied gas, be provided with metallic conductors, or electrodes, sealed into its ends, a current of electricity may be made to pass through the gas by connecting these to a suitable source of potential-difference. The stream of electricity does not simply pass directly from one electrode to the other, but part of it strikes the sides of the glass tube. Where it does so, it causes a luminous phosphorescent appearance on the glass. The stream appears to proceed from the low potential electrode—the Cathode. It passes in straight lines, as can be shown by a special form of tube having an obstacle placed in the path of the stream. Behind this obstacle no phosphorescence is excited, it appears to throw a shadow on the glass. The effect is as though *rays* proceed in straight lines from the cathode. They are spoken of as Cathode Rays. They might consist of streams of electrified particles or of waves. Now a positively electrified body placed near the stream appears to attract it. It deflects the stream towards itself, as is evident from the fact that the phosphorescence on the glass is shifted. A negatively electrified body repels the stream. It would appear, then, that the rays really consist of negatively electrified particles, as these would be acted upon in the manner described above. Again, the rays may actually be caught by a conveniently placed conductor, which is thereby found to become negatively charged. This could not be the case if the stream consisted merely of wave-motion.

FIG. 12.

Conduction through Gases

Cathode Rays

We conclude then that the stream of electricity in the rarefied gas consists of rapidly moving negatively charged particles proceeding from the cathode. They are much smaller than the smallest of the chemists' atoms. The justification for this statement must be deferred for the present.

Electrons

The particles are called Electrons. They do not depend on the kind of gas used in the tube, nor upon the material of which the electrodes are made. They are apparently all alike and all carry the same charge. Positively charged particles are also present and can be detected by suitable means. They move in the opposite direction, and are not small like the negative electrons, but consist of charged atoms or groups of atoms. The streams of these are called Positive Rays. Positive electrons, if such things exist, have not yet been discovered.

Since electrons can be obtained with tubes of different gases, and with different electrodes, it appears that—supposing them to be derived from the gas or the electrodes—the electrons must be constituents of different kinds of atoms. Such an idea leads to many speculations as to the structure of the atom, for the atom can no longer be looked upon as the smallest portion of matter. It is indeed certain that it has a complicated structure, probably consisting of some kind of positive nucleus with one or more negative electrons revolving rapidly around it. Different chemical atoms on this view will have different numbers and arrangements of electrons. In its ordinary state an atom is not electrified; this can be accounted for on the supposition that positive and negative are present in equal quantities, and so neutralize each other. If, however, an atom loses one or more electrons it becomes positively charged, while if it gains electrons it becomes negatively charged. Charged atoms or groups of atoms free to move about in gases or liquids are called Ions.

Structure of the Atom

Imagine a vessel of air amongst which ions, some positive and some negative, are mingled. They, like the air particles proper, will all be in rapid motion. Now suppose a very small drop of water or oil to be introduced at the top of the vessel. It will slowly fall, and may in its descent "catch" one or more ions. This will cause the drop to become charged with electricity. The magnitude and sign of the charge will depend on the number and sign of the ions caught, and also on the number of electrons which the different ions have themselves gained or lost. For instance, if the uncharged drop catches a negative ion, itself the

Ions

Millikan's Experiment

carrier of one surplus electron, the charge on the drop will be negative and equal to that of one electron. If then a positive ion deficient in three electrons becomes attached to the drop, the total charge on the latter will be positive, and of magnitude equal to twice that of a single electron. All possible charges on the drop will thus be multiples of the charge of a single electron, whether the sign be positive or negative. If a positively electrified plate be placed in the vessel near the top and a negatively electrified plate near the bottom, the charged drop, between the two, will be attracted by the one and repelled by the other. Suppose the drop to be negatively charged; then, although its weight tends to cause it to fall, the electrical forces tend to cause it to rise. The drop can be observed by means of a microscope, and the change in its velocity due to the sudden attachment of a single ion detected. Every time an ion is caught the motion of the drop alters. If the ion be positive, the downward velocity of the drop will be increased or its upward velocity diminished. If it be negative, the contrary effect will be observed. It was shown by Millikan,* who devised and executed these remarkable experiments, that all the charges were simple multiples of the least charge caught by the drop. This least charge must have been that due to a single electron. The experiments thus furnish a convincing confirmation of the theory outlined above.

Charge on an Electron

Not only did these "ionic charges" consist of definite multiples of the charge of a single electron, but any charge on the drop caused by friction (due to the spraying by which it was produced) was also found to consist of an exact multiple of the same, thus showing that frictional electricity is also due to a surplus or deficiency, as the case may be, of electrons.

Frictional Charges

The charge of a single electron is represented by e, and thus, according to the electron theory, the magnitude of any possible charge could be written $\pm ne$, where n is a whole number. The charge e, of course, is exceedingly minute, and the step-by-step nature of the process of charging a body is not detectable except in special cases, such as those of Millikan's experiments.

We can now form some kind of mental picture of the state

* Robert Andrews Millikan, Professor of Physics in the University of Chicago. The first determinations similar to the above were made in 1909.

of a charged body. Consider, for instance, the terminals of the battery, Fig. 10. We look upon A as being deficient in so many electrons, and on B as having a surplus. In fact, the battery acts as a sort of electron-pump, which pumps electrons out of A and into B. It does not follow of course that the actual individuals which leave A find their way to B —we must not force such an analogy too far. Now an electron on B will tend to be repelled by B and attracted by A, and so, if a convenient path were opened from B to A, we should expect a stream of electrons to pass along it. Provided the pump continued to drain A and supply B the "current" would be continuous. Such a path is provided when A and B are joined by a conducting wire. Neglecting the mode of working of the battery, and considering only what goes on in the conductor, we are now prepared to understand the modern theory of metallic conduction.

Theory of Conduction

We have already seen that in a solid body, such as a metal, the particles or molecules vibrate about certain points from which they never move far. We must now suppose that, in the case of conducting substances, some of the electrons belonging to these molecules are free to move about and collide, just like particles of gas, in the space between the molecules proper. These take part in the thermal agitation which goes on in the substance, and the hotter the substance the faster will their irregular motion be. If one end of a copper rod be raised in temperature, these electrons, by their motions and collisions, will be chiefly instrumental in passing on the heat along the rod. They, in fact, account for the high thermal conductivity of such a metal. If, instead of setting up a temperature-difference between the ends of the rod, we set up a potential-difference, say by placing the rod across A B in Fig. 10, the free electrons in the rod will be urged along in the direction B A. Each one will pursue a zigzag path, alterations in direction and velocity taking place at each collision, but on the whole it will drift towards A. It has, in fact, a general motion in that direction superposed on its irregular thermal motions. More electrons will pass in from B so long as the battery continues to supply them. This stream of electrons constitutes the electric current. It travels in the direction opposite to that which we have been accustomed to call the direction of the current; the electrons flow from places of low to places of high potential. There is no contradiction in this, the terms high and low were applied to potential long before anything was known about electrons, otherwise these terms might have been used differently, and the charge on an electron called positive instead of negative. [In other classes of conduction, electricity may be carried by positive ions; these travel in the "direction of the current."]

The current (using the word to denote the quantity of electricity flowing per second) will depend simply on the number of electrons which pass in that time, each one carrying its own charge. The electric current in a wire, then, is a kind of drift of electrons, superposed on their irregular heat motions.

It is clear that this added motion will increase on the whole the violence of the collisions which constantly take place. Let us neglect the thermal motions, and consider only the added effect due to the drift of electrons under the influence of the applied potential-difference, that is, to the current of electricity. Of course we shall have to make assumptions to simplify our ideas, just as we had to do in the case of the kinetic theory of gases. Suppose, then, that an electron has just collided and given up its kinetic energy to a molecule in the metal. This is the part of its kinetic energy due to its drift; we are neglecting the effect of the thermal motions. The electron begins to move again, under the action of what we may now call the Electric Force acting upon it, and, just as in the case of a body falling from high to low level under the action of gravity, its velocity continually increases until another collision takes place and its kinetic energy is again given up to the metal. Suppose its velocity at the moment of impact to be $2v$ and its mass m, then the energy given up will be $\frac{1}{2} m (2v)^2$, or $2 m v^2$, if the electron be brought quite to rest. This energy will appear as heat in the metal, as we should expect when we remember our study of the kinetic theory of gases.

Heating Effect

Now, on the average, the velocity of the drifting electrons will be v, if our typical electron is a fair sample, and therefore, assuming a given number of electrons per c.c. in the metal, the heat generated per second in the metal will be proportional to the square of this velocity, for it is proportional to $2mv^2$, which is the same thing as saying (since $2m$ is constant) that it is proportional v^2.

Now the current is proportional to the number of electrons passing per second, and this is proportional to the average velocity v, according to our assumption of constant electron-density (that is, the number of electrons per c.c. constant).

Thus, current $\propto v$, and heat generated $\propto v^2$. So that we arrive at the conclusion that heat will be developed in the conductor due to the passage of the current, and that for a given conductor the quantity of heat generated per second will be proportional to the square of the current.* If Q represents the heat generated per second and C the current, or quantity of electricity passing per second, then

$$Q \propto C^2.$$

The more the electrons are impeded in their drift the greater will be the development of heat if they are forced along (by the application of greater potential difference to the ends of the wire) at the same rate as before. But this means that the resistance of the wire is greater than before, for, since current is unaltered and potential-difference is greater, we see from the expression $F \propto \frac{D}{R}$ (where F can now be replaced by C) that R must be greater. In fact, for a given current, it can be shown experimentally that

$$Q \propto R.$$

Combining this with our previous result, we may write

$$Q \propto C^2 R,$$

which applies whatever the resistance of the wire and whatever the current.

The resistance of a wire depends not only on the material of which it is made, but also on its length and sectional area. If its length be doubled, the potential difference between the two ends being unaltered, the fall of potential per unit length is only half as great as before, and this means that the force acting on an electron is halved, and consequently its velocity of drift will be halved (assuming the time interval between collisions to be unchanged), and, therefore, the current will be halved. That is, in the equation $C = \frac{D}{R}$ we have kept D constant and halved C by doubling the length of the wire, thus R must be doubled. In fact, the resistance of a wire of given material and section is proportional to its length. If, on the other hand, we keep to the original length and double the sectional area, keeping D also constant, we have evidently twice as many electrons available for carrying electricity.

The Resistance

* The law of the generation of heat in the electric circuit is known as Joule's Law. It was discovered experimentally by Joule, about 1843.

Their drift velocity is the same as before, and therefore we obtain double the current. Thus in $C = \frac{D}{R}$ we have D unaltered and C doubled, thus R must be halved. In fact, the resistance of a wire of given material and length is inversely proportional to its sectional area. Thus, for a given material, the resistance is proportional to the length and inversely proportional to the sectional area. Since the passage of an electric current through a wire involves a continuous generation of heat, it is clear, if we remember the principle of conservation of energy, that energy must be expended in order to drive the current. If the current does no other work, the energy necessary to drive it is exactly equal to the heat developed. Suppose we take such a case; we know that the heat developed is proportional to $C^2 R$. By a correct choice of units we are enabled to write

The Energy

$$Q = C^2 R$$

where Q denotes either the heat developed per second or the energy necessary to drive the current C through the conductor of resistance R, when no other work is being done by the current. Now Q must be ergs, and then an appropriate unit for C chosen. We shall see later what this unit is. The unit of resistance will then be equal to the resistance of a conductor in which one erg of heat energy is developed during the passage of unit current for one second. These units are called the Electromagnetic units of current and resistance. By Ohm's law, $C = \frac{D}{R}$, we see that

Choice of Units

$$Q = CD \text{ and } Q = \frac{D^2}{R}$$

are alternative expressions for the energy. In these forms D, of course, is the potential-difference between the ends of the wire measured in electromagnetic units. From $C = \frac{D}{R}$ we may say that this unit is equal to the potential-difference which will cause unit current to flow along a wire of unit resistance, since if we make C and R each equal to one unit in the formula, D must be equal to one unit also.

We have adopted the electromagnetic system of units in the above discussion. It is the system which obtained in our formula for the attraction between magnetic poles in air :

$$F = \frac{mm_1}{r^2}$$

But we noted, in the formula

$$F = \frac{ee_1}{R^2}$$

for charged particles, that the electrostatic system of units was applicable. Now, if we wish to express current C in terms of the charged particles which give rise to it, and simply write $C = ne$ where n denotes the number of electrons flowing per second and e the charge on an electron, it is obviously necessary to express current and charge in the same system of units. If C is to be in electromagnetic units as above, we must express charge in these units also. We will write

$$C = n\varepsilon$$

and consider ε to represent the charge on an electron in electromagnetic units. The electromagnetic unit of charge is about 3×10^{10} times as great as the electrostatic unit. Thus ε is numerically much smaller than e, as it represents the same quantity of electricity in much larger units.

CHAPTER VII

Electrolysis—Laws of Electrolysis—Electromagnetic Effects—Force on an Electron—Determination of $\frac{e}{m}$—Mass of an Electron.

CERTAIN solutions conduct electricity. Let two copper plates be attached to the terminals of a battery and placed in a glass vessel containing a solution of copper sulphate in water; a current will flow through the solution. Although a molecule of $Cu\,SO_4$ is electrically neutral, its two parts, if dissociated in the solution, Cu and SO_4, are positively and negatively charged respectively. They are called Ions.

Electrolysis

On account of the difference of potential between the two copper plates, or Electrodes, such ions will travel through the liquid, the positive ones (Cu) towards the negative electrode or Cathode, and the negative (SO_4) towards the positive electrode or Anode. Both these streams help to carry the current from anode to cathode in the solution. When the positive ions reach the cathode they give up their charge and stick to the plate; thus metallic Cu is deposited on this plate. Incidentally, this is the principle of electroplating. The SO_4 ions also give up their negative charge to the anode and, uniting with it, reform $Cu\,SO_4$. Thus the anode gradually dissolves in the solution. The ions thus play a part similar to that of the electrons in metallic conduction; but here ions of both signs assist in the process. The solution is called an Electrolyte, and the process we have outlined Electrolysis.

The mass of any substance liberated from solution, like the Cu in the above case, depends on the nature of the substance and on the quantity of electricity passing. Each Cu ion carries a definite quantity of electricity, and each SO_4 ion an equal quantity of opposite sign (for $Cu\,SO_4$ is neutral). It is evident that, since the same number of Cu and SO_4 ions take part in the process, the mass of Cu concerned is to the mass of SO_4 as the atomic wt. of Cu is to the atomic wt. of $S + 4$ times that of O. So, for a given current, definite proportions of the two will be liberated. Also it is evident that the quantity of Cu will be proportional to the quantity of electricity passing, i.e., to current $\times$ time, for current is quantity of electricity passing per second. Suppose now H_2SO_4 to be substituted for $CuSO_4$. The negative SO_4 carries the same charge as before, and since H_2SO_4 is electrically neutral, it follows that the two atoms of H in the solution carry an equal positive charge to that previously carried by the Cu ion. Thus, given the same current as before, the mass of hydrogen liberated will be to the mass of copper liberated in the last case as twice the atomic weight of hydrogen is to the atomic weight of copper. In fact, the masses of different substances liberated or decomposed by a given current are in the proportions of their chemical equivalents, and as we have just seen they are also proportional to the current itself. These results were discovered experimentally by Faraday.*

Laws of Electrolysis

One of the most important effects of the electric current still remains to be described. Suppose A B, Fig. 13, to represent a wire carrying a current *C* which flows from A to B, and P to be a magnetic pole. This may be supposed to be at the end of a long thin magnet, the other pole being very distant. The effect referred to is a force, due to the current, which acts on P, and tends to move it in a direction perpendicular to A B and also to P N, where P N is a perpendicular from P on A B. This direction, in the case figured, is perpendicular to the paper at P. If P be a positive pole, the direction will be from the reader, and if negative towards him. If P were free to move continuously, it would circulate round and round A B in the path shown dotted, for its motion must always keep perpendicular to P N, which continually changes its direction as P moves. This circle is called a Line of (magnetic) Force. A magnetic pole placed at any point of it is urged in the direction of the curve at that point. Similar circular lines of force may be imagined everywhere surrounding A B so long as the current *C* flows. The direction moved by the positive pole would be

Electro-magnetic Effects

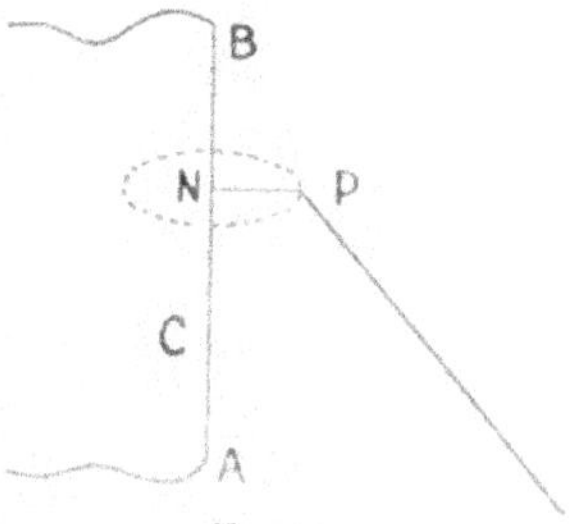

FIG. 13.

* Michael Faraday, 1791–1867. A distinguished chemist and natural philosopher. His great work, "Experimental Researches in Electricity," covered a very extensive field.

reversed if the direction of C were reversed. The actual force on the pole depends on three factors. It is (1) proportional to the strength of the magnetic pole; (2) proportional to the current; and (3) inversely proportional to the distance of the pole from the wire A B. If we suppose the pole P to be fixed, and A B to be capable of motion, though always remaining parallel to its original direction, it will tend to travel around P in a circular path. In Fig. 13, P being positive, A B would tend to start by moving bodily out of the paper towards the reader, the direction of the current being from A to B.

These phenomena show that magnetism and electricity are intimately connected with one another. The region around a magnetic pole, or around a wire or other medium carrying an electric current, is called a Magnetic Field. It may be mapped out by means of lines of force, each line showing the direction of the magnetic force (or force which would be exerted on a unit positive pole) at each point of its length. This force is a measure of the strength of the field, or the Magnetic Intensity at the point. In the case illustrated in Fig. 13 the field due to the current C is a variable one; in certain cases, however, it may be uniform; then the lines of force will be parallel straight lines.

Similarly, the region round a charged body is called an Electric Field. It is mapped out by lines of electric force which indicate the direction of the Electric Intensity or force which would act on a unit of positive electricity at each point. Every line starts from a positive charge, and ends on a negative one.

We will consider a case of special interest which bears on the electron theory. In Fig. 13 we have two magnetic fields superposed, one due to C and one to P. These give rise to a resultant field. We have seen that P being fixed and A B free to move, the latter will move towards the reader. Now the field due to P, an isolated positive pole, is represented by a set of straight lines radiating from P in all directions, since a unit positive pole situated near to P would be repelled from P in a straight line. One line of force due to P must be P N, and therefore at N the direction of the field (to adopt an expression which is often used) due to P is along that line, and towards the left. The tendency of the bit of wire near N is to move in a direction perpendicular to the field and also perpendicular to the direction of the current, i.e., in a direction perpendicular to the plane of the paper. The same applies to every other part of A B, the only difference being that while the direction is the same for every part, the intensity of the force acting on the wire gets less and less as the distance from P increases. If the wire A B were placed in a uniform field instead of a variable one, the force on every part of it would be the same. To sum up, we may say that every part of a conductor, carrying a current, placed in a magnetic field tends to move in a direction perpendicular to the field and also to the current itself. If these be at right angles to one another the force acting on unit length of the conductor is equal to the product of the magnetic field (say H) and the current.

Force on an Electron

From the point of view of the electron theory we may suppose the force to act on each electron which is acting as a carrier of current. Let there be **n** of these per unit length of the conductor, each carrying a charge ε and moving with velocity v. The number passing per second will be $\mathbf{n}v$ and the charge passing per second $\mathbf{n}\, v\, \varepsilon$. This is the current, so we have:

$$C = \mathbf{n}\, \varepsilon\, v.$$

The force on unit length is $H\, C$; this acts on the **n** electrons in the unit length, i.e., on **n** electrons we have the force $H\, \mathbf{n}\, \varepsilon\, v$ and consequently on one electron the force is $H\, \varepsilon\, v$.

Determination of $\frac{e}{m}$

Now to come to our special case. Instead of a conductor carrying a current, imagine a stream of cathode rays or negatively charged particles in a tube. Let a magnetic field be set up so that the lines of force pass through the tube at right angles to the stream, as shown in Fig. 14, where A B represents the original direction of the particles, and H that of the magnetic field. The force F acts on each particle at right angles to the stream and to H. Its direction (the slight variation of which, due to the change in the direction of the particles, we will neglect) is as shown. The stream will be deflected by F and will eventually reach C instead of B. This point will be indicated by a phosphorescent patch on

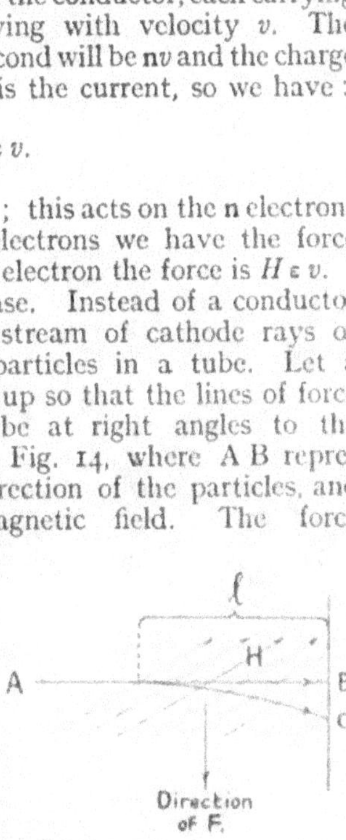

FIG. 14.

the wall of the tube, or by a spot on a photographic plate placed to receive it. Let distance B C $= x$. Suppose each charged particle to have mass m and charge ε, and suppose it to be moving with velocity v. Now the force F due to the magnetic field acts on the particle and is, as we have seen, equal to $H \varepsilon v$. But if a force F acts on a particle of mass m for time t, the distance moved in the direction of F (the particle being originally at rest and free to move) is equal to $\frac{1}{2}at^2$, where a is the acceleration $\left(\text{i.e., } \frac{\text{force}}{\text{mass}}\right)$, thus the distance $= \frac{1}{2}\frac{F}{m}t^2$. (The fact that the particle is already moving at right angles to the force makes no difference to the result; the two motions at right angles to one another may be looked upon as quite independent of each other in this respect.) Substituting the value for F obtained above, we have the distance in direction of F given by

$$x = \frac{1}{2} H \frac{\varepsilon}{m} v t^2.$$

If l represents the length of path of particle over which the magnetic force acts, as shown by the diagram, we have $l = vt$ or $t = \frac{l}{v}$, and therefore:

$$x = \frac{1}{2} H \frac{\varepsilon}{m} \frac{l^2}{v}.$$

The quantities in this equation supposed known or measurable, in an actual experiment, are x, H and l. If v were also known, the value of $\frac{\varepsilon}{m}$ could be determined; this is the ratio of charge to mass for a single electron. But v cannot be determined from this equation; it is therefore necessary to extend the investigation.

Suppose a uniform electric field to be set up at right angles to the magnetic one on Fig. 14, so as to coincide in direction with the arrow F. Each negative electron will experience a force, say f, acting along this field, but in the direction opposite to F. The magnitude of the force will be equal to the product of the field intensity, say E, and the charge ε of the electron, for E is the force acting on unit charge. In fact $f = E\varepsilon$. If this force acted alone, that is, without the opposing magnetic force F, the electrons would be urged upward in the figure and the distance y above B, at which they would strike the screen, would be given by an expression similar to the one for x, so that $y = \frac{1}{2}\frac{f}{m}t^2$. Here, however, $f = E\varepsilon$, and therefore

$$y = \frac{1}{2}E\frac{\varepsilon}{m}t^2.$$

This differs from the expression for x (apart from the fact that E replaces H) in that it does not contain the quantity v; the electric force, unlike the magnetic force, being independent of the velocity of the electron. If now two experiments be performed, one with H and the other with E, these quantities being supposed known, and the values of x and y measured, v can be obtained, for

$$\frac{x}{y} = \frac{\frac{1}{2}H\frac{\varepsilon}{m}vt^2}{\frac{1}{2}E\frac{\varepsilon}{m}t^2}$$

$$= \frac{H}{E}v.$$

Practically it is better to perform both experiments, as it were, in one, and so arrange the strengths of the fields H and E as to make x and y equal to each other. Then, since the electrons are being urged equally in opposite directions, the result is that no deflection at all takes place. This condition can be obtained by so adjusting the fields that the spot of light remains at B. Now, since $x = y$, we have

$$1 = \frac{H}{E}v \quad \text{or} \quad v = \frac{E}{H}.$$

By taking a new experiment with H only, and measuring x and then substituting the value just determined for v in the equation

$$x = \frac{1}{2}H\frac{\varepsilon}{m}\frac{l^2}{v}$$

we obtain, supposing H to have been the same for both experiments,

$$x = \frac{1}{2}\frac{H^2l^2}{E}\frac{\varepsilon}{m}$$

or

$$\frac{\varepsilon}{m} = \frac{2E}{H^2l^2}x,$$

from which $\frac{\varepsilon}{m}$, the ratio of the charge on an electron to its mass, can be obtained.

These experiments were devised and carried out by Sir J. J. Thomson in 1897. Assuming that the charge e (now expressed in electrostatic units again) is the same as that carried by hydrogen as an electrolytic ion, for which the value of $\frac{\text{charge}}{\text{mass}}$ was already known, his earliest results indicated that the mass of the cathode particle was about 770 times less than that of the lightest known atom—the atom of hydrogen.* Subsequently, as we have already seen (this sketch is quite unchronological), Millikan determined the value of e, and thus that of m can be obtained. Using a more recent value for $\frac{e}{m}$, namely, 5.3×10^{17} and Millikan's value, 4.77×10^{-10} for e, we obtain for the mass of an electron $m = 9 \times 10^{-28}$ grms. The mass of the hydrogen atom is about 1.66×10^{-24} grms., and therefore is about 1,800 times that of an electron. Now the charge carried in electrolysis by 1 gram of hydrogen has long been known to be about 2.89×10^{14} electrostatic units of electricity, so that carried by one hydrogen atom must be $2.89 \times 10^{14} \times 1.66 \times 10^{-24}$, i.e., 4.8×10^{-10} units, which is the same as that found by Millikan for the charge on an electron.

Mass of an Electron

Millikan's determinations of e were not the earliest, several others, giving rather less consistent results, having been made in 1897 and subsequent years at the Cavendish Laboratory by other methods.

Determinations of the ratio of charge to mass for positive rays show that these consist of charged atoms or groups of atoms.

* "All preconceived notions he sets at defiance
By means of some neat and ingenious appliance,
By which he discovers a new law of science
Which no one had ever suspected before.
All the chemists went off into fits;
Some of them thought they were losing their wits,
When quite without warning
(Their theories scorning)
The atom one morning
He broke into bits.

"*Here's a health to Professor J. J.*
May he hunt ions for many a day.
And take observations
And work out equations
And find the relations
Which forces obey."

A. A. R.

CHAPTER VIII

Unit of Current—Force at the Centre—Practical Applications—Induced Currents.

THE lines of magnetic force due to a current in a long straight conductor form circles around it. If the conductor be bent into the form of a circle, these lines become merged and distorted. In Fig. 15 some of these

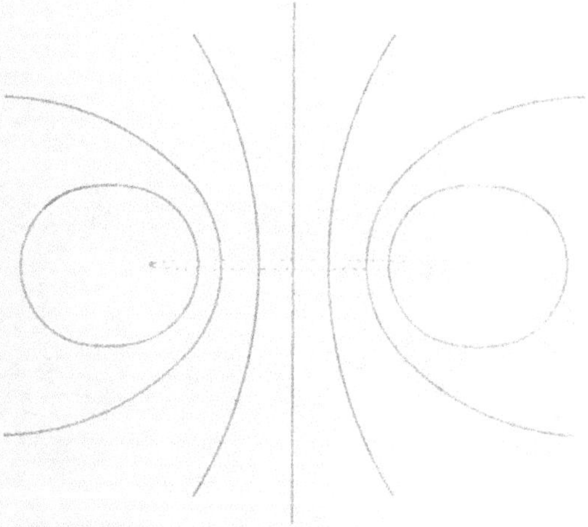

FIG. 15.

lines are shown. The magnetic force is greater where the lines are more closely concentrated than where they are widely separated. One line in this diagram is straight. A unit magnetic pole placed at the centre would be urged along this line, with a force decreasing as the distance from the coil increased. Suppose the radius of the coil to be 1 cm. and let the current be so adjusted as to give a force of 2π dynes on the unit pole at the centre. In this case, since the length of the circular conductor is 2π cms., each cm. of the conductor carrying the current is responsible for a force of 1 dyne on the pole. Under these circumstances the current is called **unit current.** This unit is the electromagnetic unit of

Unit of Current

current. It is the unit which is used in scientific calculations. The practical unit, the Ampere, is equal to $\frac{1}{10}$th of this.

Force at the Centre

The force at the centre of a circular coil of n turns, of radius *r*, carrying a current *C* (in electromagnetic units), is equal to $2\pi \times n\frac{1}{r}C$ (in dynes) on a unit pole, or $\frac{2\pi n \ Cm}{r}$ on a pole of strength *m*.

The disposition of lines of force due to a solenoid or coil similar to that shown in Fig. 16, is, so far as the field

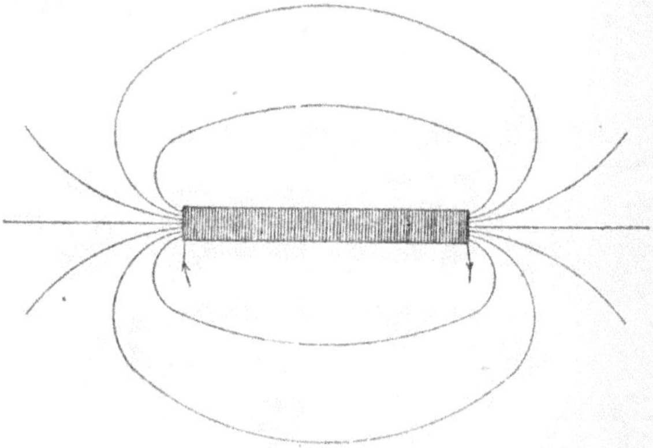

FIG. 16.

outside the coil is concerned, like that due to a bar magnet. Inside the solenoid the lines are nearly straight. The number of lines, and consequently the intensity, may be greatly increased by filling the solenoid with soft iron, which becomes strongly magnetized by the field in which it is situated while the current is flowing in the solenoid. Such an arrangement constitutes an Electro magnet.

We have seen that a conductor, carrying a current placed in a magnetic field, perpendicular to the direction of the field, tends to move in a direction perpendicular to the field and to its own length. An arrangement depending on this fact is shown in Fig. 17. The coil A B C D, through which a current can be passed, is supported by a fine wire or metal strip capable of being twisted, the amount of twist being proportional to the forces applied to the coil. The lines of force due to the permanent magnet N S, whose ends are bent towards each other, are perpendicular to the parts A B and C D of the coil in its initial position. When a current passes in the direction shown by the arrows, A B is urged towards the reader and C D away from him. This causes a twist of the suspended system—which may carry a light pointer—the degree of twist depending on the strength of the current. The deflection of the pointer thus furnishes an indication of the strength of the current, the whole arrangement forming a Galvanometer of the "moving coil" type. The instrument must be calibrated by passing known currents through it, and observing the corresponding deflections of the pointer. The maximum deflection will be 90°, for evidently A B and C D cannot move further in a direction perpendicular to the plane of the paper when this deflection is reached. Theoretically an infinitely strong current would be required to produce this deflection. Practically the instrument is only used with currents which give comparatively small deflections.

Practical Applications

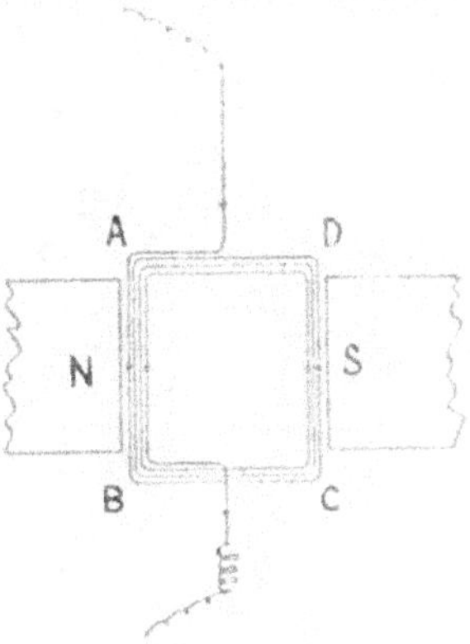

FIG. 17.

The Electric Motor depends on exactly the same principle as the moving coil galvanometer, but means must be adopted so as to give continuous rotation to the coils employed, so long as the current is supplied. Another electromagnetic phenomenon, allied with those we have been considering, remains to be discussed. Fig. 18 represents a wire A B in a magnetic field *H*. Suppose A B to be moved bodily in a direction perpendi-

Induced Currents

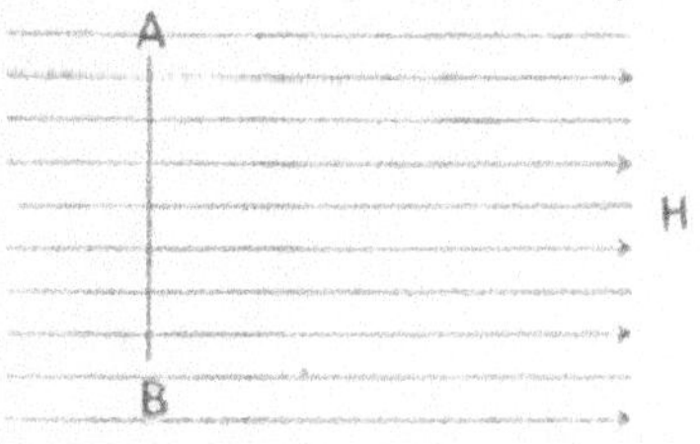

FIG. 18.

cular to its length and to *H*, i.e., towards, or from, the reader. During the motion, so long as A B remains in the field *H* (which is supposed to extend some distance above and below the plane of the paper) a difference of potential is found to exist between the extremities A and B. If these be joined by another wire suitably arranged, say a long loop which passes out of the range of H, a current will pass. This will stop when the motion ceases. It is called an induced current, and is proportional to the strength of field *H* and to the velocity of motion of A B. In this way electric currents can be generated by mechanical means. The field must be produced by a permanent magnet or an electro-magnet; in the latter case the necessary current is supplied by the Dynamo, as the machine is called, itself. A small field due to permanent magnetism must exist in order to allow the dynamo to begin its work of generating current, then the electromagnet gradually gains strength and the current increases to its maximum value. The current produced is never quite uniform; it may pass always in one direction, or its direction may alternate rapidly, according to the construction of the coils and connexions.

CHAPTER IX

The Field—Radiation—Hertzian Waves—Wave Motion—Reflection and Refraction—Wave Length—Interference—Young's Experiment—Polarization—Sound—Maxwell's Theory—Electromagnetic Waves—The Spectrum—Diffraction—Dispersion—Construction of Huygens—Colour.

A REGION of space may, as we have seen, be the seat of electric and of magnetic force. The magnitude may vary from point to point and from moment to moment. Electric force may be due to an electric charge, fixed or moving, or to a magnetic charge in motion. Magnetic force may be due to a magnetic charge, fixed or moving, or to an electric charge in motion. Suppose an electron or other small charged body to be motionless in space. Lines of electric force radiate from it in all directions. The force is everywhere directed towards the body if its charge be negative. The magnitude of the force varies with distance according to the inverse square law. An electric field exists around the body, but no magnetic force is anywhere present. Now let the body begin to oscillate rapidly to and fro along a short straight line. Evidently a disturbance will be set up in the electric field close to the body. In a short time the effect will be felt at a greater distance, and as time goes on the field farther and farther from the oscillating body will become the seat of a rapidly varying force, the variations keeping time with the motions of the body, though always late by the amount of time which the disturbance took to travel to the point in question. In fact, waves of electric force spread out in all directions from the oscillating charged body. These waves take time to travel. Their speed is about 3×10^{10} cms. per second.

The Field

Radiation

But since the charge is now in motion, magnetic forces also appear in the field. Such a field is called an Electromagnetic Field. The forces vary in value, keeping time with the movements of the body. One second after the beginning of the oscillation, the magnetic force will be felt at a distance of 3×10^{10} cms., and at the same instant the first disturbance of the original electric field will there take place. So long as the body oscillates—and indeed for a second after it has ceased to do so—that region of space will be the seat of oscillating electric and magnetic forces, or of electromagnetic oscillations. The whole field around the source—the oscillating body—is filled with electromagnetic waves, which travel outward, like ripples on a sheet of water. Lines along which the waves progress, straight lines in the present case, may be called rays, and the body a source of electromagnetic radiation. When such radiation falls on electrons, free to move, the forces act upon them and set them into motion. If the electrons are the free electrons of a suitably placed conductor alternating currents may be set up. The radiation, then, can be detected by suitable means. Waves of this type were first demonstrated by Hertz* in 1887, who used as source a rapidly alternating discharge, or current, passing across an air space between two metallic terminals. Wireless telegraphy is based upon these experiments.

Hertzian Waves

Rays of the type we have been considering are fundamentally different from cathode rays. The latter consist of streams of charged particles. The charge may be caught and measured. But electromagnetic rays carry no charge. They do, however, convey energy, as is evident from the fact that currents, which could be made to do work, may be produced in conductors on which they fall.

Certain types of radiation are familiar to us from common experience. Light and so-called Radiant Heat are the chief. Sound may be classed as radiation—it proceeds from a source and spreads with diminution of intensity, like the others. Is light analogous with cathode or with electromagnetic radiation? That is, does it consist of streams of particles shot out by the luminous source, or of waves? The former, according to Newton. But the electro-magnetic researches of Maxwell† and the experiments of Hertz had not then been accomplished. Let us look more closely into this subject of wave motion.

Waves are of many kinds. We will consider two fundamental types—longitudinal and transverse waves. In Fig. 19A

Wave Motion

the point o represents the source of a longitudinal wave travelling along o d, supposed perpendicular to the paper. The lines a a′, b b′ are drawn to

* Heinrich Hertz, German Physicist, 1857–1894.

† James Clerk Maxwell, 1831–1879. Professor of Physics at Cambridge, and one of the greatest of modern natural philosophers. He founded the electromagnetic theory of light about the year 1865.

help the perspective. Figs. 19B and 19C represent transverse waves. In each case the dots represent successive portions of the medium through which the wave progresses. Each of these portions moves backwards and forwards over a short range as the wave passes. The motion of each lags a little on that of the one behind it. In case A the track of each dot consists of a short length of o d. In B each track is parallel to a a′, and in C to b b′. These cases can very easily be illustrated by means of a cord fixed at one end and agitated by the hand at the other. Every part of the cord follows the motions of the hand, at least approximately, until the wave reaches the fixed end. The wave consists of a series of crests and troughs which follow one another with uniform

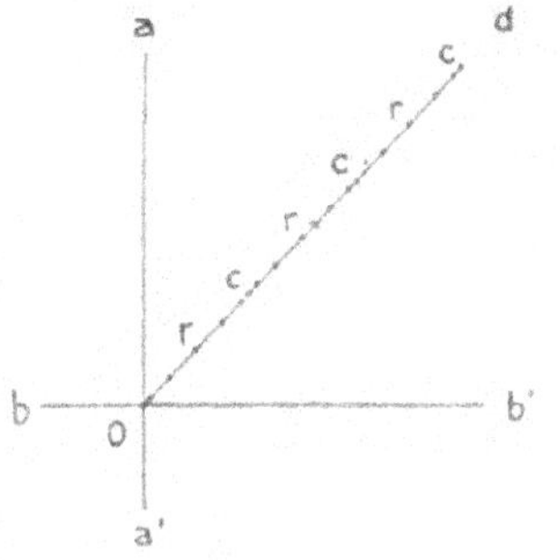

FIG. 19A.

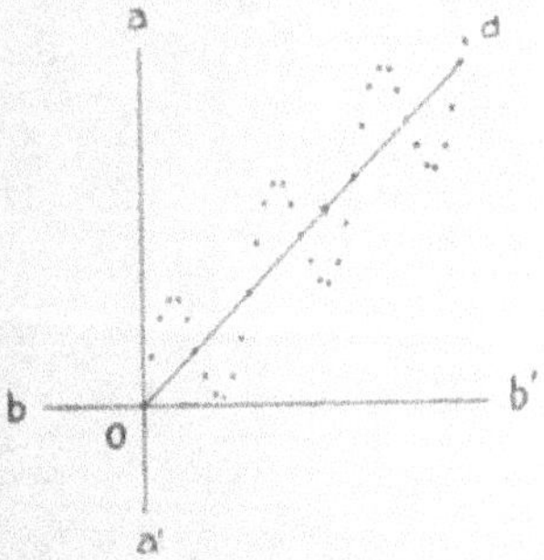

FIG. 19B.

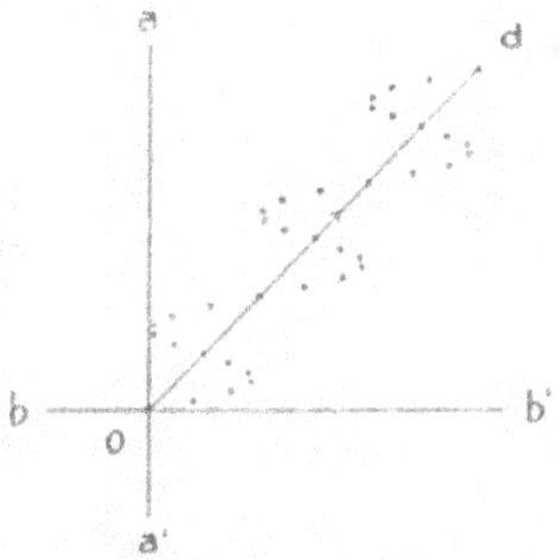

FIG. 19C.

velocity. Case A is not quite so easy to illustrate; it is usually done by means of a long spiral spring suspended by threads. One end of the spring may be moved to and fro in the direction of its length, and the successive compressions and rarefactions, which replace the crests and troughs of the transverse waves, can be watched as they travel along the spring. These are marked c and r in Fig. 19A. Every wave travels with a definite velocity: the greater the lag between successive portions of the medium the less the velocity of the wave. Some media will transmit both kinds of waves—sea-waves are transverse (though not purely so), but water also transmits longitudinal waves, the velocity of the latter being much the greater.

Reflection and Refraction

Waves in a homogeneous medium travel on without deviation. If two media join at a plane or curved surface, a wave in one of them may be reflected on reaching the surface. If, however, the second medium be suitable for transmitting the particular type of wave concerned, part of the energy of the wave may travel on through the second medium; in general the direction will be altered. This phenomenon is known as Refraction. If the wave-transmitting properties of a medium vary gradually from point to point, the path of a wave through it will in general be curved.

Wave Length

Waves are produced by vibration of the medium which carries them. A rod dipped vertically into a sheet of water, and moved up and down **n** times per second, will act as a source of waves which will travel out in circles with a velocity depending on the particular properties of the water which are concerned in the carrying of waves of this type. The frequency of the vibrations is n. The distance from crest to crest or from trough to trough of the waves is called the Wave Length. Let it be represented by λ. Since **n** of these waves are generated in one second, the distance moved by a given wave in a second must be $n\lambda$; thus if V be the velocity we have:

$$V = n\lambda.$$

Interference

When a train of waves strikes a smooth, impenetrable surface, reflection occurs. Fig. 20 shows a case of reflection of circular water ripples by a flat, solid plate. Rays may be imagined as shown by the dotted lines. But if the plate obstructs only part of the wave, as in Fig. 21 for instance—the medium extending beyond—very complicated effects occur. Take a case in which waves from a point source S, Fig. 22, fall upon an obstacle having two small openings A B. The waves reaching A and B will agitate the medium, and cause A and B to act

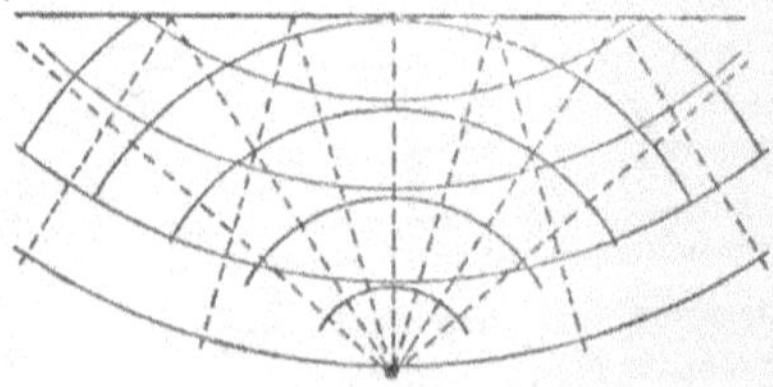

FIG. 20.

as secondary sources for waves in the medium beyond. As S is symmetrical with respect to A B, these new waves will be in the same " phase " as one another—a crest at A will be generated at exactly the same time as a crest at B. The frequency of the vibrations at A and B will, of course, be the same as that at S. Now consider the effect on the far side of A B. At P, a point equidistant from A and B, crests arrive at the same instant from both A and B. Half a vibration-period later a trough from each is due, and so on. The result will be an intensified oscillation at P. Next consider the effect at Q, where A Q is half a wave length

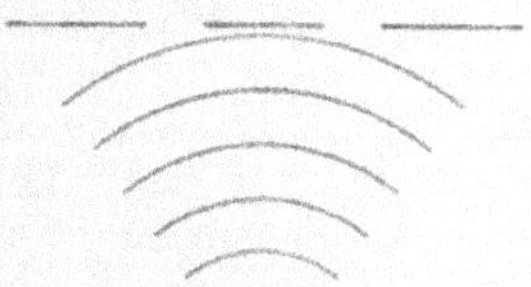

FIG. 21.

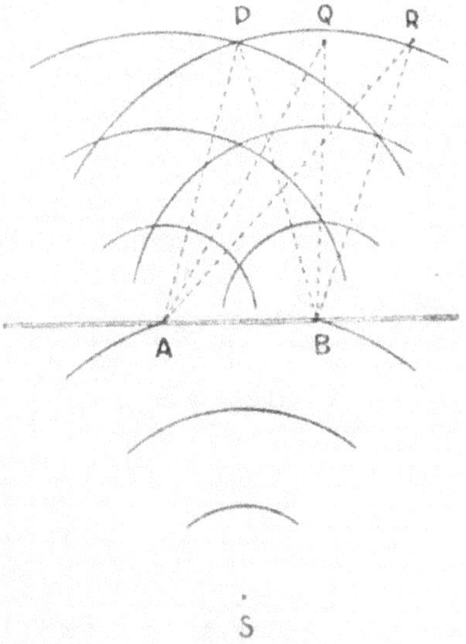

FIG. 22.

longer than B Q. At the instant of arrival of a crest from B a trough is due from A. Destructive Interference takes place—the medium cannot be moulded into a crest and a trough at the same place and time, so it remains at rest, like a body originally at rest, acted on by two equal and contrary forces. Half a period later a crest from A and a trough from B interfere, and still no motion is produced at Q. The same is true for every other instant. The intensity of the radiation is zero at Q. It is true that we have neglected the fact that the impulses from A are rather less than those from B, since

B is nearer to Q than is A. But in the main our investigation is true. At a point R, where R A is a complete wave length greater than R B, crests and troughs arrive in unison from A and B, and an intensified motion occurs. Thus on the far side of A B we have regions of maximum and minimum intensity.

If light radiation be due to wave motion, we might expect to obtain effects of the above kind by means of a suitably planned experiment. On the other hand, if light consisted of particles, like cathode rays, shot out from S, illumination could only be expected along lines directed from S through the openings A and B. The experiment was first made by Young,* who, at the beginning of the nineteenth century, obtained a result exactly corresponding to our wave investigation. The illumination of a screen placed parallel to A B was variable; it consisted of a central bright band flanked by uniformly spaced bands on either side.

Young's Experiment

Referring to B and C, Fig. 19, we may note that in the case of transverse waves a difference in the direction of vibration of the particles of the medium may exist in waves which are otherwise identical. The excursions of the particles may be parallel to a a′ or to b b′ or to any line in their plane, the direction of the wave itself still being o d. It is clear from Fig. 19, A, that no such variety can exist in the case of longitudinal waves. Suppose wave B to be one travelling along a cord o d, and let a plate, having a narrow slit cut in it, through which the cord is threaded, be held between o and d in a plane parallel to a o b, with the slit vertical. The waves are not interfered with by the plate. The plate may be said to be "transparent" to the wave. But let the plate be rotated in its own plane till the slit is horizontal. Now evidently the wave cannot pass. The plate is "opaque." But an exactly similar wave, except as regards direction of motion of the particles of the cord, such as C, could now pass through the plate. If a wave like A can pass through a plate, no rotation of the latter similar to that described above can cause an obstruction to the wave.

Imagine, instead of one cord o d, a large number of parallel cords, each carrying a wave. These may vibrate in planes making many different angles with o a. A plate with a slit for each wave, every slit being, say, vertical, would allow

* Thomas Young, 1773–1829. Professor of Natural Philosophy at the Royal Institution, London.

all the waves which happened to correspond to B to pass without obstruction. All waves like C would be completely stopped, while those vibrating at an angle with o a would be able to transmit a part of their motion. After transit these would form waves similar to B but having smaller "amplitude"—the amplitude being the distance from the line o d to the extreme limit of the excursion of a particle, or the height of a wave crest from the wave's centre-line. The waves after passing the plate, then, are all of the class B.

Polarization

They are said to be Plane Polarized. If the plate be rotated in its own plane another set will pass unmolested, and components parallel to them of others will also pass—the waves will still be plane polarized, but the Plane of Polarization will now be different from what it was in the first case. This plane is represented by b o d for wave B and by a o d for wave C. It is the plane containing the line along which the wave travels and perpendicular to the direction of vibration of the particles.

Suppose now we take a second plate exactly similar to the one described above, and place this parallel to the first a little farther along the wave, the cords threading it just as they do the first. If the position of the second plate be similar to that of the first, i.e., the slits of both being parallel, then the polarized waves which have passed the first are unobstructed by the second. If now the second be turned 90° in its own plane every wave is stopped short. On rotating it another 90° the waves pass as before. At intermediate positions, components of the waves will be allowed to pass, so that the effect of rotation will be a gradual and not a sudden extinction of the waves.* The first plate may be called a Polarizer and the second an Analyser. An exactly analogous effect can be obtained with light—I had almost said with light waves, for it must be fairly clear by now that light consists of waves and not of projected particles. A plate of tourmaline, cut in a certain way—parallel to the axis of the crystal—acts on a beam of ordinary light in the same manner as our first plate acted on the vibrating strings. It polarizes the light. To show this we have only to take a second tourmaline and hold it a little farther along the beam of light and rotate it in its own plane. At two opposite positions the plate allows the light to pass, but cuts it off at two other positions, 90°

* Strictly, of course, each little slit ought to turn about its own centre; if the plate turned as a whole the mass of strings would be twisted.

to the first. Since, then, this analyser causes an alteration in the intensity of the transmitted light from full to zero by mere rotation such as we have described, it is evident that the light falling on it has a kind of two-sided nature—it is polarized. This points to the fact that light waves are transverse and not longitudinal.

On the other hand, no such phenomenon as polarization is known in connexion with sound. But interference does take place in the case of sound. Just as two waves of light, as from A and B in Fig. 22, interfere at Q, with darkness as a result, so two sounds may interfere with each other, the result being silence. When two organ pipes of nearly equal pitch sound together "beats" are produced. These can be explained if we assume sound to be transmitted in the form of waves. If two waves reach the ear in the same phase at a given instant, the result is intensified sound, i.e., a sound louder than would be due to either wave separately. But if a little later on the waves reach the ear in opposite phase, they will interfere, and the result will be minimum intensity of sound, or even silence. Later, when the waves are in the same phase, loud sound is again heard, and so on. To accomplish this result it is necessary for the wave length of the one train of waves to be slightly greater than that of the other. Then, since the two trains travel with the same velocity, it is clear that an ear will receive the waves "in step" at one instant, and "out of step" at a succeeding instant, then in step again and so on. The listener, in fact, will hear beats. It will not be difficult for the reader to prove for himself that the frequency of the beats will be equal to the difference between the frequencies of the two sound vibrations. If, then, sound consists of waves, but of waves which do not exhibit polarization, we must identify the waves with those of type A, Fig. 19. The waves of sound are longitudinal.

Sound

It can easily be shown by experiment that sound will not pass through a vacuum. It is readily transmitted by solids, liquids, and gases. Light on the other hand passes even more readily through a vacuum than through even the most transparent substance. Different media, then, are necessary for the two classes of waves.

The wave motion constituting sound is more easily studied experimentally than that of light. The waves can be produced in air by means of a body such as a tuning fork set into mechanical vibration. The frequency of the vibrations can be determined, and the pitch of the note produced shown to depend on the frequency. Doubling the frequency raises the pitch by an octave. The amplitude of the waves (amplitude = ½ length of excursion of a particle of air) determines the intensity of the sound; and an analysis of the "wave form" shows that this controls the quality of the note. The wave form depends on the particular way in which a particle of the medium moves in its path. It might, for instance, move uniformly from end to end and back, or it might move rapidly forward and slowly back, or in any other imaginable way, the only necessary feature of its motion being that it must be periodic, that is, the same motion must be repeated exactly over and over again. Thus we see that the three qualities by means of which we can draw up the specification of any sound waves—frequency, or its allied wave length, amplitude, and form—have their corresponding effects on the sound heard, namely, on the pitch, loudness, and quality.

The velocity of sound in dry air at 0° C is about 33,130 cms. per sec. That of light, as determined by experiment, is about 3×10^{10} cms. per sec. A calculation made by Maxwell of the velocity to be expected in the case of electromagnetic waves led to this same value. This fact forms the foundation of Maxwell's Electromagnetic Theory of Light. It is simply that light waves are electromagnetic waves, and not mere mechanical ripples like those produced by agitating still water. Later, Hertz actually demonstrated the production of electromagnetic waves, as we have already seen. These waves, according to the theory, are similar to light waves, and differ only in that their wave lengths are enormously greater than those of light. This theory holds the field to-day, but it is not possible to form a mental picture of electromagnetic waves, as it was of the mechanical vibrations of the older theory—the "elastic solid theory," which supposed the vibrations to take place in a kind of elastic substance called Ether, which permeated all space. The whole question of the ether of space is far too difficult to enter on here.

Maxwell's Theory

We may take a rough survey of the electromagnetic waves of which we have knowledge. The Hertzian waves may have wave lengths ranging from 1 or more kilometres down to 1 cm. or even less. Shorter waves of lengths from about .01 cm. to less than .00001 cm. can be detected in radiations from various sources, including hot bodies and flames,

Electromagnetic Waves

and gases rendered luminous by the passage of the electric current. Of these waves those ranging approximately from .000077 to .000039 cm. excite vision if they fall on the eye. All the waves carry energy.

The group of waves emitted by a body whose temperature is below that of incandescence contains none of the shorter ones, but as its temperature is raised, shorter and shorter waves are generated, till at last the longest of those capable of exciting vision are given off. These give rise to the sensation of red. As the temperature is further raised, shorter waves mingle with the others, and the resulting light becomes whiter and whiter. Then, finally, rays too short to cause vision begin to be emitted. These, together with the shorter visual waves, give rise to pronounced chemical effects: for example, they powerfully affect photographic plates.

The Spectrum

But, in order to investigate these waves, leaving aside the Hertzian waves for the present, it is necessary to adopt means of separating out the different constituents contained in a composite beam, such as that emitted by, say, an electric arc lamp. Two methods are in common use, one depending on the interference phenomenon illustrated in Fig. 22. It is evident that the point R of great intensity will be nearer to P the shorter the wave length of the radiation, for A R is just one wave length greater than B R. So a measurement of P R enables us, together with a knowledge of the other dimensions of the figure, to calculate the wave length of the radiation emitted by S. If this radiation contains waves of several lengths, there will be a different point R for each length; the different wave lengths will, as it were, be ranged along P R. Such an arrangement of the different constituents of a beam of radiation is called a Spectrum. The spectrum we have considered is a first order spectrum; it is due to a difference of one wave length between the lengths A R and B R. There will be second and third order spectra due to differences of two and three wave lengths, and so on. Of course, also, a corresponding set of spectra will exist on the other side of P. In actual practice, a plate A B, having a very large number of parallel transparent (or reflecting) strips, separated by opaque ones, is used. These Diffraction Gratings, as they are called, have thousands of such strips or "rulings" to the inch. The investigation necessary is somewhat more complicated than in the above simple case, but is the same in principle.

Diffraction

Dispersion

The other method of sorting the wave lengths is due to the fact that waves of different lengths are differently refracted when they enter or leave (obliquely) a transparent substance, such as glass. Generally, the longer the wave the less the refraction. In Fig. 23 I P represents a composite wave incident at P on a glass block. P R_1 indicates the path in the glass of the longest waves and P R_2 that of the shortest. This separation is called Dispersion. A more convenient method is to use a prism of glass or other transparent substance (see Fig. 24). The

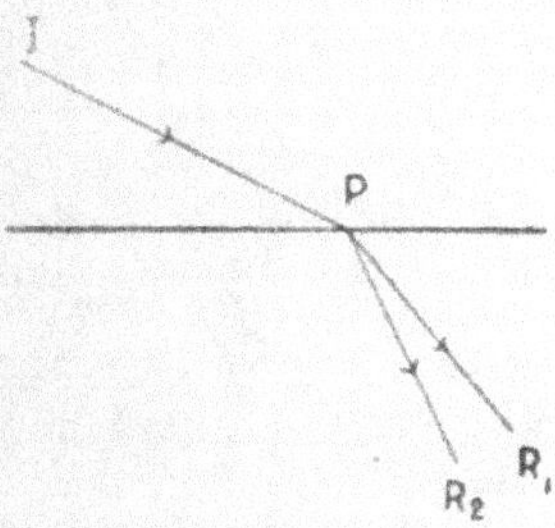

FIG. 23.

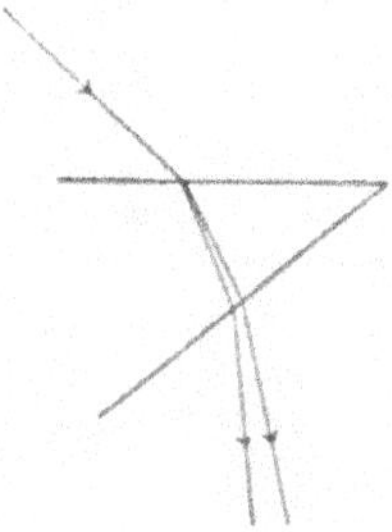

FIG. 24.

resulting spectrum may be projected on to a screen, or received on a photographic plate, or the energy carried by each portion may be determined by means of a delicate heat-measuring apparatus. The spectrum given by a prism is not quite the same as that due to the diffraction grating, the arrangement of the various wave lengths being less regular and sometimes altogether anomalous.

The fact that waves of different wave lengths are differently refracted is closely connected with another fact—that they travel in refracting substances with different velocities, although all have the same velocity in vacuo. The connexion between the two phenomena

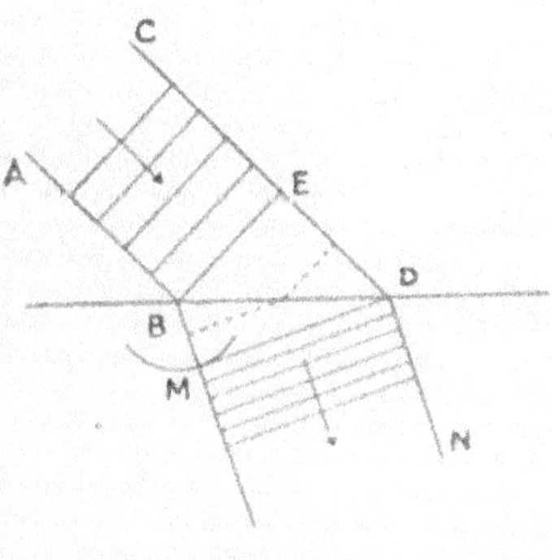

FIG. 25.

can be seen by the construction of Huygens.* Let A B, C D, Fig. 25, represent the extreme rays of a beam of radiation falling on a glass block. The lines B E, etc., represent wave fronts advancing to the surface. When the end B of a wave front reaches the surface, E has still a distance E D to travel. Meanwhile, B passes into the glass, but moves with reduced velocity. Thus when E arrives at D, B reaches some point on the circular arc, whose radius is to E D as velocity in glass is to velocity in air, for the wave concerned. So the wave front in the glass must stretch from D to the arc, to which it will be tangential at M. The new ray from B will be B M, perpendicular to D M, and that from D will be D N, which must also be in the same direction. The new wave fronts will be closer together than before; in fact, the wave length in the glass is less than that in air. This is evident from the fact that B M contains the same number of waves as E D. These intermediate wave fronts will be bent where they cross the surface B D; one of them is indicated by the dotted line. It is clear that the amount of bending or refraction of A B and C D depends on the amount of reduction of velocity on entering the substance, for the smaller the velocity the smaller the radius B M and, consequently, the greater the refraction.

Construction of Huygens

Let us now consider a spectrum, such as that given by a diffraction grating, containing all wave lengths which have been experimentally obtained. We neglect the long Hertz waves, to which the grating is not adapted. The central portion of the spectrum yields waves which excite vision. Of these the longest give rise to a sensation of red, the shortest to that of violet. Waves of intermediate lengths correspond to the remaining colours of the rainbow. Mixtures of these waves in varying proportions give rise to all the colour sensations which we can perceive. A mixture of all in suitable proportions produces white. Other rays not in this " visible spectrum " may also be mixed in without altering the visible effect. Beyond the ends of the visible spectrum we have waves called Infra

Colour

* Christian Huygens, Dutch philosopher, 1629–1693. He propounded the undulatory theory of light, which, however, was not generally accepted till long after. He also discovered the polarization of light, but could not explain it, as he had longitudinal waves and not transverse ones in mind.

Red and Ultra Violet respectively. The most powerful heating waves are in the infra red and the red end of the visible spectrum, and those which chiefly affect the photographic plate are in the ultra violet and towards the violet end of the visible spectrum. The waves which affect the eye most strongly are those in the yellow and green portions of the visible spectrum; the effect grows weaker as either the extreme red or violet is approached.

CHAPTER X

X-Rays — Their Production — Crystals — Wave Length — Radio-activity—Radio-active Changes—The Theory—Ionization—Atomic Energy.

A NEW type of rays was discovered by Röntgen* in 1895. These rays, called by him X-rays, possess very remarkable properties. They readily penetrate considerable thicknesses of substances such as wood, aluminium, and muscular tissue, which are opaque to ordinary light. Bone is more opaque to these rays than flesh, and the heavy metals are more opaque still. Like the shorter waves of the spectrum, X-rays affect the photographic plate, and they also excite fluorescence when they fall on certain substances. On account of these properties, shadow photographs of, say, the bones of the body can be taken, or shadow pictures observed directly by the substitution of a fluorescent screen for the photographic plate. But, unlike light, X-rays cannot be refracted by prisms and lenses, or reflected by mirrors.

X-Rays

X-rays are produced when cathode rays strike a solid obstacle. They start from the point at which the collision takes place. They are quite dissimilar from the cathode rays which give rise to them, for they carry no electric charge and they cannot be deflected by a magnetic or an electric field. For the convenient production of them a stream of cathode rays is caused to impinge on a small metal plate in a discharge tube; X-rays are given off, and pass out of the tube in straight lines radiating from the metal target.

Their Production

It might seem that a corpuscular theory would best fit the positive and negative properties to which we have referred; but recent experiments by Laue and W. H. and W. L. Bragg show that effects similar to some of those exhibited by light can also be obtained with X-rays. In the Braggs' experiments a beam of X-rays is caused to fall on a plate of crystal. Owing to the regular arrangement of the atoms in the crystal certain planes exist within it which are rich in atoms, while other planes contain fewer atoms. Such

Crystals

* Wilhelm Konrad von Röntgen, German physicist.

planes are represented by A B, A C, and A D in Fig. 26, A B containing the greatest number of atoms per unit area. These planes act as reflectors to the rays. The more atoms per square cm. the more efficient the reflector. Suppose a beam of X-rays to fall obliquely on A B. A small fraction of the beam will be reflected, and the rest will pass on. Of this remainder a small portion will be reflected at the next plane of atoms parallel to A B, and further portions at the successive planes in the crystal. These reflected portions will (in certain circumstances) reinforce each other, and emerge from the crystal as a beam which can be detected by its action on a photographic plate. In order that reinforcement may occur, the direction in which the reflected rays travel must be such that the paths of the successive portions differ from one another in length by exact multiples of the wave length—we are assuming a wave theory of X-rays. In other directions there will be destructive interference between them, and consequently no emergent beam will exist. The difference of path between two of the portions depends on the distance between the successive reflecting planes, as well as on the angle of reflection, and it is therefore evident that a relation exists between the wave length of the X-rays and this distance between the layers of atoms in the crystal.

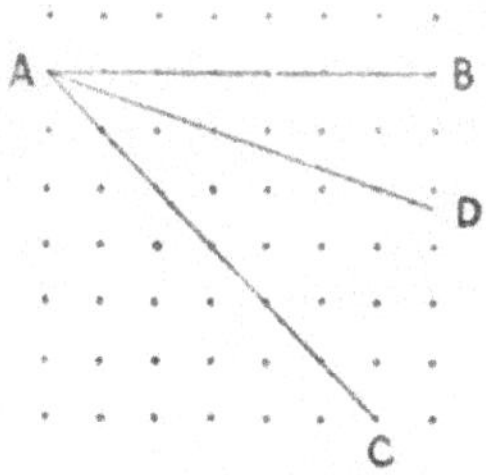

FIG. 26.

The Braggs succeeded in measuring both these quantities, and in locating accurately the atoms in many crystals. The wave lengths were found to be of the order of .00000001 cm. or about $\frac{1}{500}$ of the shortest ultra violet light waves observable by direct methods. It is due to the extreme shortness of these waves that the fine grained mirrors, naturally supplied as we have seen by crystals, are necessary for the experiments on reflection, and that ordinary mirrors are ineffective. The penetrating power of the rays is also due to the same cause. X-rays, indeed, possess the properties of very short light waves, and we may place them at the end of our scheme of electromagnetic radiation, which begins with long Hertzian waves, and includes the infra red, visual and ultra violet light.

Wave Length

We have seen that in electrical processes, such as the dis

charge through rarefied gases, some electrons become detached from the atoms of which, in ordinary circumstances, they form a part. Atoms, then, are not the indivisible entities of the older atomic theory. But we cannot, apart from such cases, artificially split them up into their component parts. A discovery, however, made soon after that of X-rays, by Becquerel, and extended by M. and Mme. Curie, shows that certain atoms may spontaneously disintegrate. The substances whose atoms exhibit this property are called Radio-active. A radio-active substance, such as a mass of radium, which has been left alone for a considerable time, emits rays of different kinds, known as α-, β-, and γ-rays. The α-rays have been identified with atoms of helium, carrying a positive electric charge equal to $2e$. The velocity with which they are emitted varies somewhat for the different radio-active substances; it is one or two times 10^9 cms. per second. The α-particles are capable of producing phosphorescent flashes on striking a suitably prepared screen. The atom ejecting such a particle thereby becomes lighter—indeed, it becomes a different atom.

Radio-activity

The β-rays emitted by radio-active substances are similar to cathode rays. They consist of electrons, with the usual negative charge of magnitude e. Their velocity is generally greater than that of cathode rays, and in some cases is very nearly equal to the velocity of light. Both α- and β-rays possess a certain power of penetrating substances, this being much the greater in the case of β-rays. α-rays can be stopped by small thicknesses of solid matter, and the fastest of them can only penetrate a few cms. of air; β-rays on the other hand penetrate several metres of air. This is due to their superior velocity; but if α- and β-rays of equal velocities were compared, the α-rays, being much heavier, would be found to be the more penetrating of the two. Since the mass of a β-particle is very small, its loss makes very little difference to the mass of the atom ejecting it.

The γ-rays are similar to X-rays, and carry no charge. They do not consist of actual particles like the other types, and are not deflected, as are the others, by electric and magnetic fields. They have much greater penetrating power than that of the β-rays.

A heavy radio-active atom such as uranium may give off successively a number of α-particles or charged helium atoms. At each ejection it becomes a new atom. Thus it passes through a whole series of changes; the final product of the series beginning with uranium has been identified with lead. β-rays are also given off, but, as we have seen, the effect of this on the mass is inconsiderable.

Radio-active Changes

The emission from a given radio-active substance, such as the mass of radium referred to above, goes on at a constant rate, which depends on its quantity only, and is not affected by its temperature or other physical conditions. The actual process of radio-activity is, however, much more complex than would appear from this simple fact. The substance can be split up by chemical means into several portions, which emit differently: some α-rays, some β- and γ-rays, and some all three kinds, though details of the process vary for different substances. One of these portions usually consists of the main bulk of the substance, the others of very small quantities. The total emission, consisting of the emissions from the several portions added together, remains constant, and equal to the original emission from the undivided substance. But the activity of each of the portions varies as time goes on, and would eventually stop altogether in the case of all except the first. This portion grows in activity as the others wane, so keeping the total activity constant.

The Theory

The results may be explained on the supposition that the separate portions each contain at first one of a series of active substances, which differ among themselves. Each of these gradually decays, but in doing so produces some of the substance next in the series, which in its turn decays and gives rise to the next kind and so on. The final product is not radio-active, at least to any observable degree. When all the portions which were originally separated from the main mass have decayed into this final form, a whole new series has been produced in the mass, and thus the total activity is preserved constant. Of course, in time the whole substance must decay and lose its activity, but its life is very long. The actual changes which go on are accompanied by the emission of rays, as we have already seen; but only an extremely minute fraction of the atoms composing a radio-active substance such as uranium or radium takes part in the changes at any given time.

The series of active substances above referred to is called a Radio-active Series. The rates at which the several constituents decay vary exceedingly. Radium decays to half its value in about 1,580 years, and other members of the series following radium in about 3.85 days, 3 minutes, 27 minutes,

20 minutes, 16 years, 4.85 days, and 136.5 days respectively.

X-rays and the rays from radio-active substances possess the power of "ionizing" gases through which they pass.

Ionization

Air in its ordinary condition is an exceedingly good insulator, but it at once becomes capable of conducting electricity if a beam of any of the above rays be passed through it. In its new state the air is said to be Ionized. The rays, falling on gaseous atoms, detach electrons from them, and thus the originally neutral atoms become divided into positively charged atoms on the one hand and negative electrons on the other. Such a gas in an electric field conducts electricity, the positive and negative ions, which move in opposite directions under the action of the field, giving rise to a current through the gas.

Atomic Energy

From what has been said it will be clear that the atoms of matter contain considerable stores of energy. Some of this energy is expended whenever radio-active changes occur. If means could be discovered by which it could be extracted and used at will it would be found to exceed enormously in quantity the energy which is developed in the ordinary processes of chemical action and combustion. Indeed, a beginning has already been made, for Sir E. Rutherford, by bombarding nitrogen with α-particles, has succeeded in breaking up this element into hydrogen and a gas of mass 3, with emission of rays whose energy is greater than that used in their production; so that we may be on the verge of the discovery of new available sources of energy.

PART II

PART II

FORCE AND ACCELERATION

TWO preliminary experiments:

(*a*) *Verification of the Principle of the Triangle of Forces.* In this experiment three forces acting at a point are represented by strings attached to a small metal ring, and kept in a state of tension by means of three stretched spring balances hooked to pegs fixed in a drawing board. The magnitudes of the forces can be read on the scales of the balances. Trace the directions of the forces, that is, of the strings, on a paper pinned to the board, and mark off along each direction a distance proportional to the force represented. Verify the statement that a triangle can be drawn whose sides taken in order represent in magnitude and direction the three forces acting on the ring. Also show that the resultant of any two of the forces—obtained by drawing the diagonal of the parallelogram (shown in Fig. 27) of which the two lines, representing the forces, from adjacent sides—is equal in magnitude and opposite in direction to the third force. Alter the position of one of the pegs and repeat the experiment.

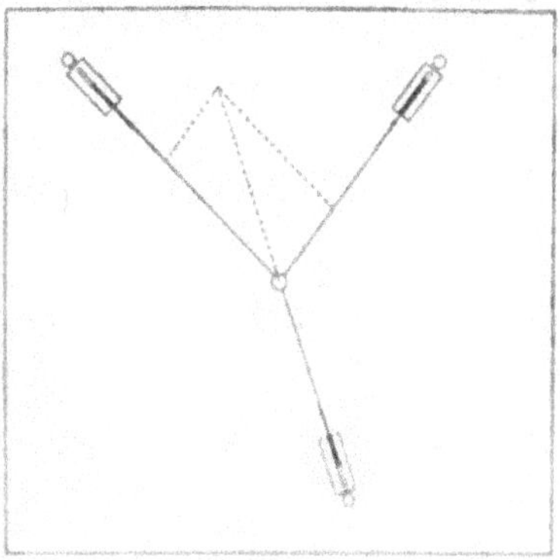

FIG. 27.

(*b*) *Determination of g, the Acceleration of Gravity, by the Simple Pendulum.* Set up a simple pendulum by suspending a metal ball, a cm. or two in diameter, from a fixed support by means of a fine thread. Draw a line on a card, and fix it just behind the pendulum, with the line vertical, so that, when viewed from a point in front, the pendulum in its position of rest appears to coincide with the line. Set the pendulum swinging through a small angle and observe it from a distance of one or two metres. Time the pendulum by observing twenty, or for greater accuracy fifty, complete oscillations. Each time it crosses the central mark a " transit " is said to take place. A complete oscillation is the movement which occurs between a transit in a given direction and the next transit in the same direction. Time by means of a stop watch, and divide the total time by the number of oscillations observed, starting and stopping the watch at the exact instants of the initial and final transits. Let the time in seconds of an oscillation be t. Measure the length l of the pendulum, from the point of support to the centre of the ball. Repeat the observations, using several different lengths of thread.

The relation between t and l is given by the equation

$$t = 2\pi \sqrt{\frac{l}{g}},$$

where g is the acceleration of gravity. Since

$$g = 4\pi^2 \frac{l}{t^2},$$

as is evident by squaring the equation, it follows that $\frac{l}{t^2}$ must be constant, however l may vary, for g is constant. Verify this by calculating $\frac{l}{t^2}$ for each length used. Experimental errors will cause slight variations of the value. Take the mean value and use it to calculate the value of g by means of the equation.

The expression for t given above applies only to small amplitudes of oscillation, the amplitude being the angle between the central position of the pendulum and the extreme position on either side. Make a few observations of time of swing, using large amplitudes. Compare these times with that obtained for the same length of pendulum vibrating through a small angle.

Experiments with an Inclined Plane.

A small trolley, with light, freely moving wheels, and a smooth board, which can be set at any small angle with the horizontal, are required. The latter forms a track along which the trolley can run. Means must also be provided for determining the times taken by the trolley to travel different distances down the track. A plan sometimes adopted is to fix a strip of paper or card on the trolley, and to arrange a vibrating steel spring, fixed at one end and carrying an inked brush at the other, in such a way that the brush traces a curved line on the paper as the trolley descends the plane. The time of vibration of the spring is known, and the distance moved during one vibration is equal to the distance between two wave crests on the curve traced. If the speed of the trolley accelerates, the distances between the successive crests will increase, each distance being described in the same time. Otherwise a metronome may be set to give suitable beats (these being timed by means of a stop-watch), and the distance moved by the trolley observed while the beats are counted.

The arrangements for timing being completed, set the plane at a small angle to the horizontal, and allow the trolley to start from the rest near the top at a given instant. Note the time t taken for it to move any definite distance, say s, in cms., down the plane. Assuming that the body moves with uniformly increasing velocity, determine the final velocity at the distance s from the starting point. To do this make use of the fact that the final velocity is twice the average velocity, which is itself equal to $\frac{s}{t}$. Thus the final velocity v is equal to $\frac{2s}{t}$.

Next calculate the acceleration, f say, of the motion down the plane, that is, the change of velocity during a given time divided by that time. Thus $f = \frac{2s \div t}{t} = \frac{2s}{t^2}$, in cms. per sec. per sec. The force acting on the body in the direction of its motion is equal to the mass of the body multiplied by the acceleration. Let m be the mass of the trolley, then (ignoring a small correction due to the rotation of the wheels, whose mass should be small compared to that of the rest of the body) the force in dynes will be mf, or $\frac{2ms}{t^2}$. The actual value of this force can therefore be obtained from the data in hand.

Now calculate the force by another method. Suppose the

angle made by the plane with the horizontal to be $\theta°$, shown (exaggerated) in Fig. 28. The weight, in dynes, of the body

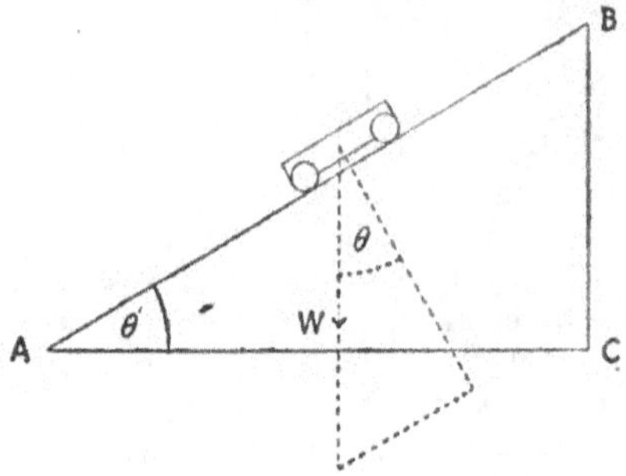

FIG. 28.

is mg, and it acts vertically downward. The component of this weight resolved parallel to the plane is alone effective in causing motion down the plane. mg being represented by the vertical barbed line, the value of the components parallel and perpendicular to the plane can be found by drawing lines in these directions so as to represent, with the vertical line, a triangle of forces. The angles of this triangle are equal to those of the triangle A B C, the angle at C being a right-angle. It is easily seen that the components parallel and perpendicular to the plane are $mg \sin \theta$ and $mg \cos \theta$ respectively. The latter is effective only in causing a pressure between the body and the plane; it takes no part in helping the motion. Sin θ can easily be determined by measuring B C and A B and taking their ratio; g has already been determined by means of the pendulum. Thus the force $mg \sin \theta$ which produces the acceleration down the plane can be determined. Compare this value with that already obtained.

It is not to be expected that the two values will exactly coincide. The second method (assuming, of course, that no error has been made in the determination of g) gives the actual force acting on the body, due to its weight, parallel to the plane. But the first method gives the mass × acceleration, and this product is equal to the above force *less* the resistance due to friction, etc. Thus the two results will only be equal if the resistance be zero. This will not be so in any actual case. Let the resistance to motion experienced by the body be denoted by F, the direction of which lies along the plane in the sense opposite to the motion. Then the force producing motion will be not $mg \sin \theta$, but $mg \sin \theta - F$, and the true equation of motion will be

$$mg \sin \theta - F = \frac{2ms}{t^2}.$$

To investigate more fully the frictional resistance, take a case in which it is more pronounced—substitute a wooden block for the trolley, and attach a light pulley to the plane, over which a thread, fastened to the block, and carrying a scale pan at its other end, may pass. Perform the following experiments on

Friction

First set the plane horizontally, and arrange block, thread, and scale pan as shown in Fig. 29. Weigh the block and the

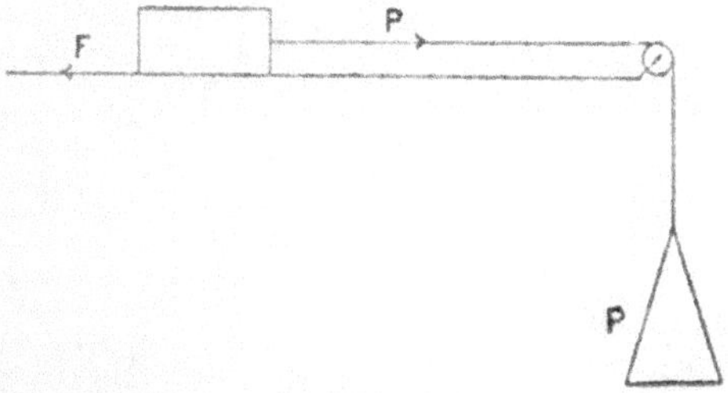

FIG. 29.

pan, and adjust weights in the latter till, on being given a start with the finger, the block moves uniformly along the plane without acceleration. The force acting on the block in the direction of motion is equal to the weight hanging by the thread; and this force, say P, must be equal to the frictional resistance F; if it were greater the velocity of the block would increase, and if it were less the block would come to rest. Thus we have $P = F$. Now, the law of friction applicable to this case states that F is proportional to the force with which the block is pressed on to the plane, which, since the plane is horizontal, is equal to W, the weight of the block. Thus $F \propto W$ or $F = \mu W$ where μ is a constant, called the Coefficient of Friction, depending on the nature of the surfaces in contact. We have then

$$P = \mu W,$$

from which the value of μ can be obtained. Add different weights to the block and readjust P in each case so as to obtain several determinations of μ.

Next set the plane to an angle θ with the horizontal, and adjust P so that, on being given a start, the body slides uniformly up the plane. Here (see Fig. 30) the component R

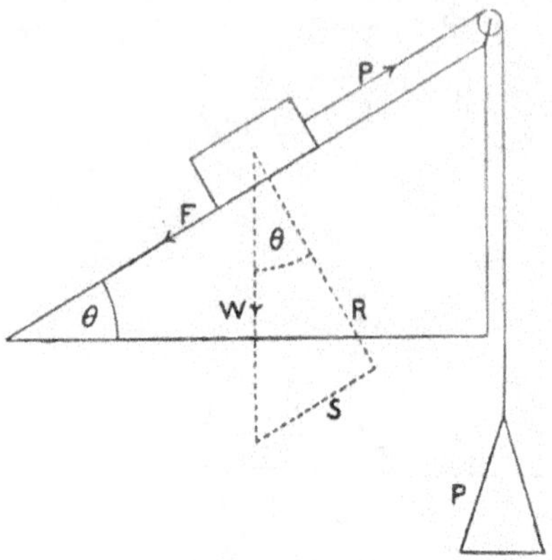

FIG. 30.

of the weight W represents the force pressing the surfaces together; thus we have $F = \mu R$ in this case. The component S acts on the block parallel to the plane, and opposes P. F also opposes P; therefore we must have $P = F + S$. But

$$R = W \cos \theta \qquad S = W \sin \theta,$$

thus the equation $F = \mu R$, or $P - S = \mu R$ becomes

$$P - W \sin \theta = \mu W \cos \theta,$$

from which μ may be obtained; $\sin \theta$ and $\cos \theta$ being determined as in the previous experiment with the inclined plane.

Lastly, remove the thread, and adjust the inclination till the block, on being given a start, slides uniformly down the plane. Here F acts in opposition to the motion, as shown in Fig. 31. The only force acting down the plane is the component S of the weight of the block. Thus we have $S = F$, that is, $W \sin \varphi = F$. But as before $F = \mu R$ or $F = \mu W \cos \varphi$, and therefore $W \sin \varphi = \mu W \cos \varphi$ or

$$\mu = \tan \varphi.$$

The fact that the weight W has disappeared from the equation indicates that the angle φ is independent of the weight. Test this by adding weights to the block and repeating the observations. φ is called the Angle of Friction.

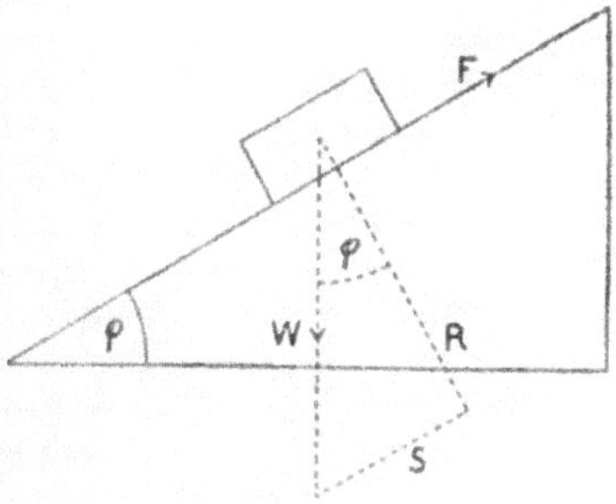

FIG. 31.

Now to return to the experiment with the trolley. Two determinations were made—of $mg \sin \theta$, and of $\frac{2ms}{t^2}$ respectively. The equation of motion was found to be

$$mg \sin \theta - F = \frac{2ms}{t^2},$$

and therefore the value of F could be obtained. It is desirable, however, to determine F experimentally; and the last experiment, on the angle of friction, indicates the method of doing this. Since $F = \mu R$, and R varies with the inclination—being equal to mg times the cosine of the angle of inclination—we must not confuse the F's of the two cases. But μ is constant, and if we find μ for the trolley by the angle of friction method, we can then find F for the original inclination θ by writing

$$F = \mu\, mg \cos \theta.$$

Thus let φ be the new inclination of the plane, such that, on being given a start, the trolley moves with uniform velocity down it. Then $\mu = \tan\varphi$, and consequently the equation

$$mg\sin\theta - F = \frac{2ms}{t^2},$$

or

$$mg\sin\theta - \mu\, mg\cos\theta = \frac{2ms}{t^2}$$

becomes

$$mg\sin\theta - \tan\varphi\, mg\cos\theta = \frac{2ms}{t^2},$$

or

$$g\,(\sin\theta - \tan\varphi\cos\theta) = \frac{2s}{t^2}.$$

Take several values of θ, and observe corresponding values of s and t, and substitute in each case in the equation. Or rearrange the equation thus:

$$g = \frac{2s/t^2}{\sin\theta - \tan\varphi\cos\theta}$$

and obtain a value for g for each set of observations of s and t. In this last form the experiment furnishes a means of determining the value of g, alternative to that of the simple pendulum.

EXPERIMENTS ON AIR

Barometer and Thermometer

The height of the barometer, as actually observed, depends not only on the pressure of the atmosphere, but on the temperature of the barometer itself. This is because the density of the mercury and also the length of the scale attached to the instrument vary with temperature. But if, the pressure remaining the same, the temperature were reduced to 0° C., the observed height would be correct, for then the mercury would be at the definite density which is always assumed in such determinations, and the scale should also be correct. Now, pressure due to a column of mercury of density ρ is $gh\rho$ (in dynes per sq. cm.). If the density were ρ_0 and the corresponding height h_0 the pressure (remaining the same) would be given by $g h_0 \rho_0$. Thus

$$h_0 \rho_0 = h \rho \text{ or } h_0 = h \frac{\rho}{\rho_0}.$$

If then ρ_0 and h_0 correspond to 0° C., and ρ and h to the actual temperature t of the barometer, it is necessary to multiply the reading h by $\frac{\rho}{\rho_0}$, which is equal to $\frac{1}{1 + mt}$ where m is the coefficient of cubical expansion of mercury. This corrects the height to the value it would have at 0° C. so far as the mercury is concerned ; but the correction for the scale is also necessary. Now the length of any part of the scale at t° C. is to the true length of that part at 0° C. as $1 + lt$ is to 1, where l is the coefficient of linear expansion of the material, probably brass, of which the scale is made. The reading must be increased in this proportion, for evidently if the scale becomes **n** times too long its reading will be **n** times too short, and must therefore be increased **n** times. If then H_0 represents the finally corrected height, we have

$$H_0 = (1 + lt)h_0$$

and therefore

$$H_0 = \frac{(1 + lt)}{(1 + mt)} h.$$

In order then to be able to obtain a correct value for H_0 we need a reliable thermometer for determining t. Most ordinary thermometers are somewhat inaccurate and require correction. The errors of the graduations 0° and 100° should be determined, the first by standing the thermometer in pounded ice moistened with distilled water for 10 or 15 minutes and noting the reading; the second by supporting the thermometer in the steam of boiling water in a steam-jacketed vessel called a hypsometer, noting the reading, and comparing this with the temperature—determined from tables—of boiling at the particular pressure of the atmosphere at the time of the experiment: this to be obtained by reading the barometer.

We are evidently in a "vicious circle," for it appears that we cannot correct the barometer till the thermometer has been corrected, and for this purpose we need the corrected reading of the barometer. Such cases sometimes occur, and the general procedure is to arrive at a sufficiently accurate result by a series of successive approximations. We might, for example, assume the thermometer reading to be approximately correct, use it to correct the barometer, use this to correct the thermometer, then use the corrected thermometer to re-correct the barometer, and so on as often as thought necessary. An extra complication would arise in this case, since the barometer would be continually varying. Fortunately, however, no such complicated process is necessary in our case. A simple numerical calculation will show that a slight error in t in the equation for the corrected height of the barometer will produce only a quite negligible error in H_0. So any fairly good thermometer may be used for the correction of the barometer reading, and the result may safely be employed in the correction of the same or another thermometer, which should then be kept and used for subsequent experiments.

A graph should be made by which true temperatures can be deduced from readings of the thermometer (Fig. 32). Suppose in melting ice the reading was −1° C. To get the correct temperature, 0° C., it is necessary to add 1° to the reading. This addition to the reading is called the correction. On the horizontal axis of the diagram, which represents readings, mark the point −1° and set up vertically the correction for that reading, i.e., set up an ordinate 1° in length to some scale. The scale for the ordinates must be larger than that used for the horizontal measurements. Also suppose the steam reading to be 99.9° while the true temperature of the steam is 99.4°. The correction to be added to the reading is −0.5°. Plot this value below the point representing the reading. Join the two points obtained by a straight line. This will show the amount to be added

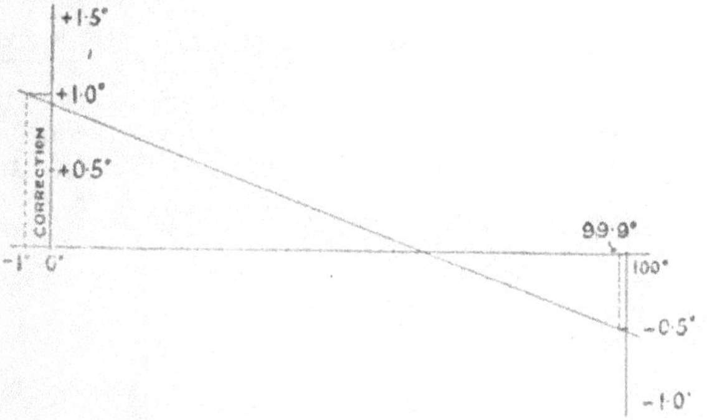

FIG. 32.

to or subtracted from any reading. It assumes that the bore of the tube is uniform and that the degree marks are equally spaced. In future, whenever the thermometer is used, always apply the necessary correction as shown by the graph. If a second thermometer be required in any experiment, compare it with the first by immersing both in the same bath of water and noting any difference between the readings at the temperature of the bath. A comparison at two temperatures may be necessary in some cases. From these a correction curve could be drawn for the second thermometer.

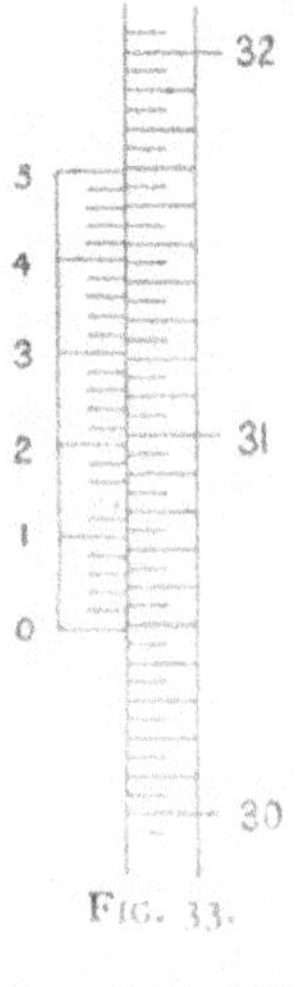

FIG. 33.

[*Note on the Vernier.* The barometer vernier will probably be of the type which reads to $\frac{1}{500}$ of an inch. Starting with the zero of the vernier at a definite inch mark, move it so as to bring the first small division (after the zero) of the vernier into coincidence with a scale division. The vernier has moved $\frac{1}{500}$ inch, for 25 such steps would involve a movement of $\frac{1}{20}$ inch, or one small scale division. This can easily be tested with the actual instrument. Now 5 small steps similar to the above would involve a motion of $\frac{5}{500}$, or .01 inch. This would bring the division marked 1 on the vernier into coincidence. The reading of the vernier in Fig. 33 is 30.494 in. The scale reading next below the vernier zero is 30.45 and the vernier reading to be added, due to the distance the vernier has moved since its zero was at 30.45, is 0.44—that is, .04 for the 4 numbered divisions, and 4 in the next place of decimals for the two odd divisions between 4 and 5. All verniers may be investigated by noting the number of small steps which it is necessary to move the vernier while its zero moves from one small division of the scale to the next. Each of these small steps corresponds to a change of the coincidence from one small vernier division to the next.]

THE GAS LAW, $\frac{PV}{T} = constant.$

P = pressure, V = volume, and T = absolute temperature of any given quantity of gas or air.

I. *Verification of Boyle's Law.* $PV = constant$ for constant temperature.

Use an apparatus such as that illustrated in Fig. 34.

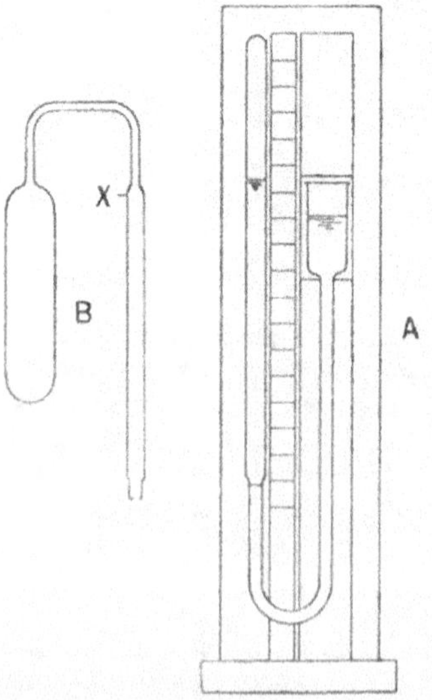

FIG. 34.

A uniform glass tube closed at one end and containing dry air is fixed vertically beside a scale, and is connected by a rubber tube to a reservoir, open to the air, which can be raised and lowered by means of a slide to which it is attached. The reservoir and the lower part of the glass tube, together with the rubber tube, contain air-free mercury. The slide is placed so that the level of the mercury in the reservoir can be read on the scale. The volume V of the air in the tube is taken to be proportional to the length, say L, of the tube which it occupies; and the pressure in cms. of mercury, which may, if desired, be corrected to 0° C., is obtained as follows. Imagine the atmosphere, which exerts a pressure on the surface of the mercury in the reservoir, to be removed, and replaced by a column of mercury equal in height to the height of the barometer. Then the pressure in the air tube is given by the difference between the level of the top of this column and the level of the mercury in the air tube.

Show that PL is constant for several positions of the reservoir, some of them giving a pressure greater than that of the atmosphere, and some less. If PL be constant, evidently PV is constant also. Variations of temperature during the experiment will affect the result, and should be avoided.

II. *Determination of the "Pressure Coefficient" for Air.* This is the coefficient of increase of pressure of air with rise of temperature, the volume being kept constant. Since V is constant, $\frac{P}{T}$ must also be constant; thus if P_0, T_0 denote pressure and absolute temperature at 0° C., and P, T at t° C.,

we have $\frac{P}{T} = \frac{P_0}{T_0}$, therefore $P = P_0 \frac{T}{T_0}$

or $$P = P_0 \frac{(T_0 + t)}{T_0}$$

therefore $$P = P_0\left(1 + \frac{1}{T_0}t\right).$$

Thus $\frac{1}{T_0}$ takes the same place in this expression for pressure as the coefficient of expansion in the expression for length or volume of a body undergoing expansion with rise of temperature. It is the pressure coefficient required. Its value is approximately $\frac{1}{273}$ for T_0; the absolute temperature of 0° C. is about 273°.

To obtain the coefficient experimentally, use an apparatus like that of Fig. 34, in which the Boyle's law tube is replaced by the bulb B with its tube. B contains dry air, and the mercury level is to be adjusted always to a mark X just below the narrowing of the tube as shown. The bulb is to be immersed in a mixture of ice and distilled water, giving a temperature of 0° C., and the position of the mercury in the reservoir observed after the adjustment at X is made. The pressure is then to be found as in the last experiment. The bulb must then be immersed in baths, well stirred, of different temperatures, and the corresponding values of P obtained. These must be plotted on a diagram to a temperature basis, absolute or centigrade, and a straight line drawn as evenly as possible among the points, as in Fig. 35. The values of P_0 and P_{100} can then be obtained from the points where this line cuts ordinates at 0° and 100°. This method tends to eliminate experimental errors. All the observations are utilized, and the straight line is a kind of compromise between them, and values of particular ordinates should be taken by reference to this line rather than to the actual points obtained experimentally. This method of using experimental results is of very wide application.

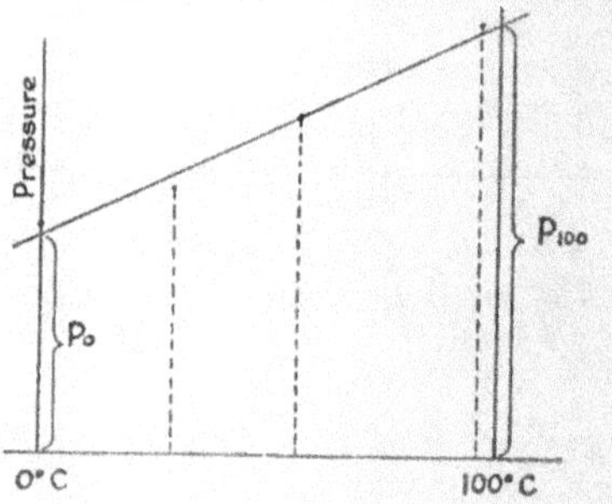

FIG. 35.

Now $P_{100} = P_0 (1 + p\ 100°)$, where p is really $\frac{1}{T_0}$ the pressure coefficient. Thus

$$p = \frac{P_{100} - P_0}{100\ P_0}.$$

The value of T_0 can, of course, be obtained, and hence the position of the absolute zero of temperature, which is T_0 degrees below 0° C. The graph could be continued towards the left, in which case the pressure line would cut the temperature axis at the absolute zero.

Constant Volume Air Thermometer

The apparatus used in the last experiment may also be used as a constant volume air thermometer. Pressures will be read as before, including pressure of the atmosphere in each case, the level of mercury being adjusted to the fixed mark. The temperature of the air in the bulb, or of the (well stirred) bath in which the bulb is immersed being t° C., we have

$$P = P_0 (1 + pt)$$

therefore

$$t = \frac{P - P_0}{P_0 p}.$$

If the apparatus is to be used as a thermometer for measuring t of the bath, it is necessary to know P_0, p, and P. Of these, p has already been found once for all, and the value will serve for any similar apparatus. But P_0 would vary from apparatus to apparatus, and for the same apparatus, if the quantity of air in the bulb were changed. P_0 should therefore be determined by using a bath of pounded ice and distilled water. Then, so long as no change is made in the apparatus, this value can be used.

Immerse the bulb in a bath of unknown temperature. Find P and calculate t. Check by testing the bath with the corrected mercury thermometer. Do this for several different arbitrary temperatures. Slight differences will, no doubt, be evident, as several small sources of error exist with which we have not dealt. The value of the checking is perhaps more apparent than real, since the mercury thermometer was employed in the experiment for determination of p. But at least it will show whether the operations have been accurately carried out. The elimination of some of the sources of error might form a suitable exercise for specially proficient students; this applies also to many of the experiments described in this book.

GROUP EXPERIMENT

IN this section we take, as an illustration of the group experiment, the determination of several properties of a solution and the variation of these with concentration. The solution may be one of common salt, and the properties chosen, Density, Surface Tension, Boiling Point, Specific Heat, and Refractive Index. Each member of the class makes the determinations for a particular concentration of the solution, and the results are finally collected and plotted, each student supplying one point for each curve. It is not at all necessary that the various determinations should be made in the same part of the session. Each member having made up a stock of solution of the required strength may make the several experiments whenever the subject concerned is dealt with in the course. They are gathered together here for convenience.

First, then, the concentrations, in grms. per litre, should be fixed so as to range from pure water (distilled, of course) to a concentration approaching saturation at ordinary temperatures. Each student should make up a large bottle full as accurately as possible. No used solution should be returned to the bottle. The temperature of the solution should be recorded whenever it is used.

DENSITY

First find the relative density of the liquid, that is, the ratio $\frac{\text{density of liquid}}{\text{density of water}}$. It may be converted into density of the liquid by multiplying by the density of water, as determined from the tables, the temperature of the experiment being noted. Now

$$\frac{\text{density of liquid}}{\text{density of water}} = \frac{\text{mass of given volume of liquid}}{\text{mass of equal volume of water}},$$

so we need to find the masses of equal volumes of liquid and water. Use the common balance—the method of using this must be learned in the laboratory.* It is an instrument for the determination of mass by comparison with a series of standard masses. In reality it balances the *weight* of a body with that of the corresponding standard masses, but while it tells us when the weights are equal, it does not tell us what either weight is. If we took the apparatus to a place where gravity exerted only half the force, so that both weights were halved, it would still balance. But since at any particular locality, say the balance room, mass is proportional to weight, and the weights are equal, it follows that the masses must be equal, and we know the mass of the standard (which is not affected by change of locality). Thus we determine the mass of a body by means of the balance. The operation, unfortunately, is termed "weighing."

"Weigh," then, a density bottle (*a*) dry, (*b*) full of the liquid, and (*c*) full of distilled water. Subtract (*a*) from (*b*) and from (*c*), and take the required ratio. Multiply this by the density of water at the temperature of the experiment.

Again, suspend a heavy solid by a fine thread from the balance; (*a*) weigh it; (*b*) allow it to dip into a beaker of the liquid; weigh; (*c*) do the same, using distilled water instead of the liquid. In (*b*) the upthrust of the liquid is a force equal to the weight of liquid displaced, so the body seems to weigh less by the weight of its volume of liquid. The standard masses which have to be removed equal the mass of this volume of liquid. So the mass of a volume of liquid equal to the volume of the solid is determined. Determine also (*a*) — (*c*), the mass of this same volume of water, and proceed as before to calculate the density of the liquid.

Surface Tension

The height to which a liquid (which wets glass) rises in a capillary tube dipped into a vessel of the liquid depends on the surface tension T, the radius of the tube r, and on the weight of unit volume of the liquid, which is ρg, where ρ = density and g = acceleration of gravity. Upward force due to surface tension = $2\pi rT$; weight of liquid supported by this force = $\pi r^2 h \rho g$, therefore

$$T = \frac{r h \rho g}{2} \text{ (in dynes per cm.).}$$

* But see Appendix B.

To find T, then, by this method, we require to know r, h, ρ, and g; g we take as 981 cms. per sec. per sec.; ρ has already been determined. The radius of the bore of the tube r must be found, and also the height h to which the liquid rises in it.

Bore of Tube. Strictly, the radius is required at the point to which the liquid rises—and the bore may not be uniform. To test this, a thread of mercury a couple of cms. long may be run into, and measured in various parts of, the tube by means of a reading microscope. (It is not worth while to describe such instruments here; their use must be learned in the laboratory.) If its length be constant, the tube may be taken as having uniform bore. The bore can be determined by measuring a long thread of mercury in the tube, and then running out and weighing it in a small weighed beaker. Since its density is known, and equals its mass ÷ volume, its volume can be obtained, that is, the volume of a known length of tube, hence radius of tube.

Rise in Tube. Attach the tube to a glass scale by a clip. Dip into a beaker of the liquid till the tube is wetted well above the place where the liquid will finally stand. Fix in a support. A little sliding pointer clipped on to the scale will give the exact level of the liquid in the beaker. It must stand well clear of the curved surface of the liquid near the scale. (See Fig. 36.) Note the scale reading of the top of the liquid column, holding the eye as nearly as possible on the same level as the top of the column. If thought necessary a card may be set up in a stand in front of the apparatus with its upper edge at nearly the same height above the bench as the column. Observe just over the edge of the card. Remove the scale and its attachments from the liquid, taking care not to alter the position of the pointer on the scale. Observe the reading of end of pointer on the scale, so placing the eye that the pointer and its reflection in the glass of the scale are in the line of sight. The correct reading will thus be obtained. If the image be not clearly visible, hold a small piece of mirror in contact with back of scale, having carefully dried the latter if necessary. Try a rough experiment first, in order to find out any difficulties which may arise. Very poor results are often obtained, the chief reason being that the slightest trace of grease on the surface of the liquid completely alters the value of T. Contamination of the tube also prevents the surface from taking up its correct position. Any grease present forms a film on the surface of the solution. If tube, scale, beaker, and bottles in which the solution was mixed or kept were not thoroughly cleaned, bad values of T are certain. At least, the T found will not be that of the solution.

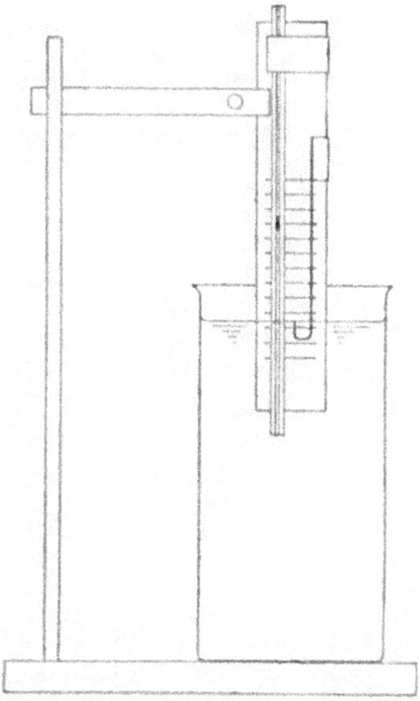

FIG. 36.

Before the experiment see that the liquid maintains its height h while the tube is raised and lowered in the solution. If it sticks, clean the tube again.

BOILING POINT

Fit a flask with cork, outlet tube, and thermometer. Take the temperature of the boiling liquid, the bulb of the thermometer being immersed in the liquid. Also take the temperature of the vapour above the liquid. Replace the liquid by distilled water, and repeat. Suppose the temperature taken in the steam over the distilled water to be 99.5° C., the barometer being rather low. A correction of .5° added to the temperature will give the true value for standard atmospheric pressure. This same correction should be added to all the other temperatures taken, and the results will be sufficiently near to the correct values for standard pressure for all practical purposes. Each of these temperatures may be plotted on the final Boiling Point diagram. As boiling proceeds, the concentration of the solution will increase, so the determinations should be taken as quickly as possible. A condenser fitted to the outlet pipe would improve matters. But in elementary work we often have to content ourselves with *knowing how* matters could be improved.

Specific Heat

I. *Method of Mixtures.* If substances at different temperature be mixed till a uniform temperature is obtained, no chemical action or change of state taking place, each substance gains or loses a number of calories equal to the product of its mass, its specific heat, and its change of temperature. Thus an equation (total calories lost = total calories gained) can be formed, having on one side all such products for the substances which lose heat, and on the other the products for those which gain heat. This assumes no loss of heat during the mixing. In practice, the substances or bodies gaining heat are generally a thin copper vessel called a calorimeter, water or other liquid contained in it, a stirrer and thermometer, the whole being at the same initial temperature. The substance losing heat is first raised to a higher temperature and then passed into the liquid in the calorimeter, and stirred up till the liquid reaches a maximum temperature. Let suffixes H, L, C, T, Q denote the heated substance, the liquid, the calorimeter, the thermometer, and the stirrer; M, S mass and specific heat; and t_1, t_2, t_3 the initial temperature of H, the initial temperature of the calorimeter and its contents, and the final temperature of the whole. Our equation written at length would be (where M_H = mass of substance H and so on):

$$M_H S_H (t_1-t_3) = M_L S_L (t_3-t_2) + M_C S_C (t_3-t_2) + M_T S_T (t_3-t_2) + M_Q S_Q (t_3-t_2)$$
$$= M_L S_L (t_3-t_2) + (M_C S_C + M_T S_T + M_Q S_Q)(t_3-t_2).$$

Although masses and temperature could easily be found, it is clear that too many unknown specific heats occur in this equation for any practical use to be made of it as it stands. If, however, in a preliminary experiment we let H and L both be water, whose specific heat is unity, we can write, using W_1 and W_2 instead of H and L.

$$M_{W_1}(t_1-t_3) = M_{W_2}(t_3-t_2) + (M_C S_C + M_T S_T + M_Q S_Q)(t_3-t_2),$$

and now the only unknown is $(M_C S_C + M_T S_T + M_Q S_Q)$, which can therefore be found. Let it be determined once for all, and called E. Note that it is a constant; it remains the same throughout all the experiments. Now our original equation may be written

$$M_H S_H (t_1-t_3) = (M_L S_L + E)(t_3-t_2).$$

If one of the two specific heats S_H or S_L be known, an experiment will furnish us with the other. Let H be a solid say, and L water, then S_H can evidently be obtained. Then this solid can be used with another liquid L in the calorimeter, and the specific heat of this determined. This method is to be used for the salt solution.

If the liquid in the calorimeter be water, we have on the right hand side of the last equation the sum of the mass of the water in the calorimeter* and the quantity E. In fact, the calorimeter, thermometer, and stirrer are equivalent in this equation to E grms. of water. For this reason E is called the water equivalent of the calorimeter, thermometer, and stirrer.

In practice, a suitable steam heater is used for solid substances, such as that shown in Fig. 37. The thermometer indicates the temperature t_1. Steam is passed till the solid is heated uniformly throughout. The lower cork is then removed, and the calorimeter brought under the heater, and the solid lowered into it by means of a thread with which it has been suspended in the heater. The contents are stirred and the maximum temperature taken. The mass of the calorimeter, empty and with liquid, and the mass of the solid must be taken before beginning the actual experiment. If the water equivalent has not been determined, look up the specific heat of copper in the tables and multiply by the mass of calorimeter and copper stirrer. The thermometer need hardly be taken into account for ordinary work. A small correction to the final temperature is necessary for very accurate work, on account of heat lost during mixing by radiation, etc. It is very small if the experiment be done expeditiously. The solid should be in the form of gauze or thin strip loosely rolled together, so that the liquid can quickly take up the heat and so give little time for loss of

* Numerically, since the specific heat of water is unity.

heat. Also the outside of the calorimeter should be polished to minimize radiation, and the calorimeter should be suspended by threads in a rather larger vessel of the same shape, for protection from draughts, etc.

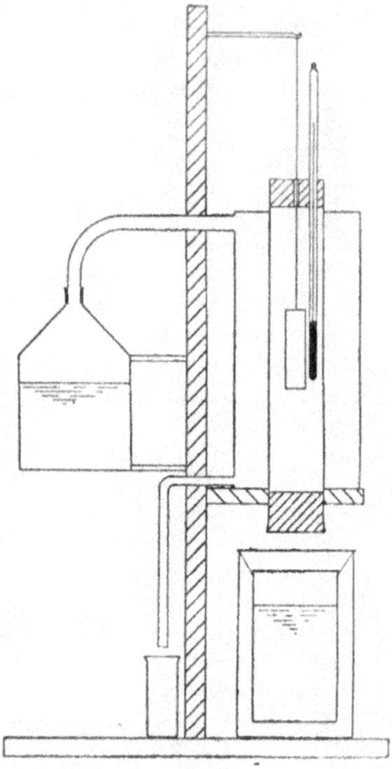

FIG. 37.

II. *Method of cooling.* Use the calorimeter of the last method in its outer case. The mass of the calorimeter (i.e., the inner vessel only) empty and about $\frac{3}{4}$ full of water at about 50° C. must be determined. (Make this determination after cooling.) A card cover should be used, through which a thermometer and stirrer pass. Keep stirring, and note temperature and time as the water cools. Plot a cooling curve, Fig. 38. Repeat, using the same volume of solution instead of water. Plot a curve on the same diagram.

Now, heat in each case is lost by radiation from the calorimeter. The heat lost per sec. depends only on the temperature of the calorimeter and of the enclosure in which it is suspended, and on the nature and area of the radiating surface of the calorimeter, not on the nature of the liquid in the calorimeter. But the total quantity of heat to be lost in falling say from 50° to 30° differs for the two cases.

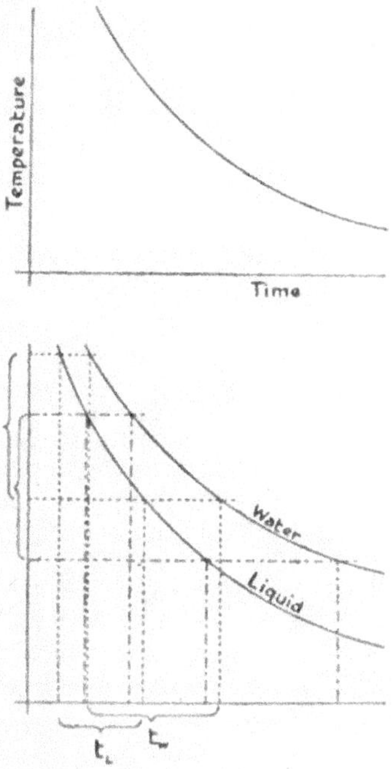

FIG. 38.

Suppose the liquid has half as many calories to lose in falling through this range as the water, it must do so in half the time, since loss of heat *per sec.* is the same at any part of the range in the two cases. The number of seconds must be half. So

$$\frac{\text{loss of heat, in case of liquid, in falling through given range}}{\text{loss of heat, in case of water, in falling through same range}}$$
$$=\frac{\text{time for liquid to fall through given range}}{\text{time for water to fall through same range}}.$$

Take the times from the curves, for several equal ranges of

temperature, e.g., the ranges indicated by brackets in the diagram. The time intervals can be read off along the time axis of the diagram for each liquid. Those for the first range are shown by brackets for water and liquid.

Now, the numerator of the left hand side of above equation is really $M_L S_L$ × (fall of temperature) + E × (fall) [the symbols having the same meanings as in last experiment], or $(M_L S_L + E)$ × (fall), and the denominator is $(M_W + E)$× (fall). If $t_L t$ be the times, we have, since "fall" cancels,

$$\frac{M_L S_L + E}{M_W + E} = \frac{t_L}{t_W}$$

whence S_L. The other ranges of temperature will give alternative values.

Refractive Index

I. *By Reading Microscope.* The refractive index μ of the liquid is equal to $\frac{\text{real depth}}{\text{apparent depth}}$ of an object in the liquid viewed from a point vertically above. Place a coin in a glass vessel, view a point on its surface by microscope, arranged vertically, with scale also vertical, as indicated diagrammatically in Fig. 39. Take the reading. Pour in the liquid, taking care not to displace the coin, and observe the image of the mark, and also of the upper surface of the liquid, which may have a few particles of fine cork dust scattered on it. From the three readings obtain the ratio for μ. One point in connexion with the use of the microscope must be mentioned—the image thrown by the object glass must coincide exactly with the cross-wires for each object viewed, otherwise the distances measured, i.e., distances through which the microscope is raised, may not be equal to the distances between the positions of the objects viewed. To ensure this, move the

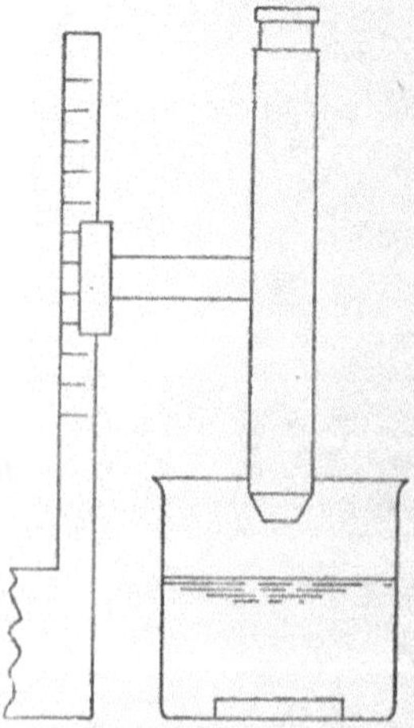

Fig. 39.

eye backwards and forwards across the eye-piece as far as possible without losing sight of the image. If image and cross-wires be not in the same plane, they will appear to move relatively to one another as the eye is moved. In this case the microscope must be raised or lowered until this relative motion does not take place. In the figure, A shows correct position of microscope and B incorrect. (The final image, due to the eye-piece of the microscope, is not shown). Several depths of liquid should be used, but note that with a very small depth experimental error will greatly affect the result.

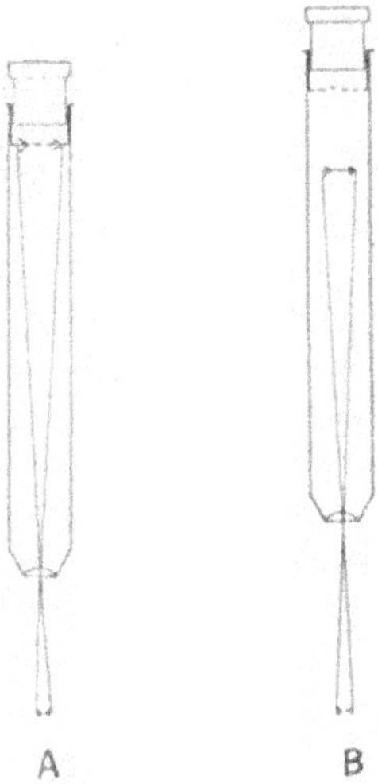

FIG. 40.

II. *By Critical Angle Method.* In this method a narrow beam of light passing through a liquid falls normally on a small air cell, composed of two plates of glass, separated by a small space, and sealed water-tight round the edges. The cell is connected to a vertical spindle, to which is attached a pointer which moves over a horizontal circular scale. Light enters at A (Fig. 41) and reaches the eye at B. The cell is now rotated till the light is completely cut off. It is totally reflected. At this point the index is read. The normal to the cell, if it exactly coincided with the beam at first, has been turned through a certain angle, say $\theta°$. But it may not

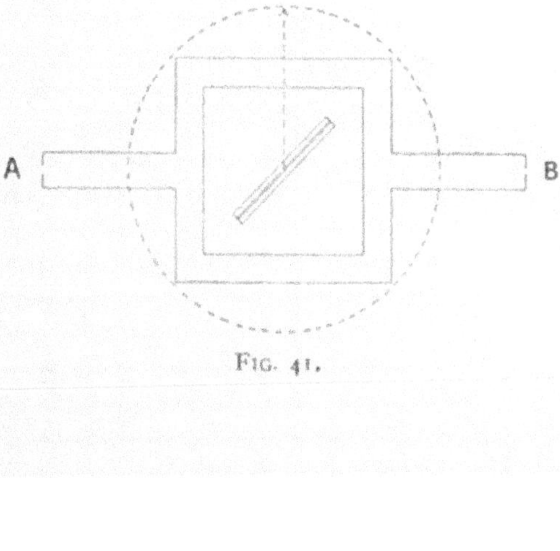

FIG. 41.

have coincided exactly—the cell was only set perpendicular to the beam by guess. But now move the cell back to its original position and through this to a position at which the light is again cut off. The normal is now $\theta°$ on the other side of the beam. Thus the total travel of the normal is exactly $2\theta°$. This is, of course, the same as the travel of the pointer, whatever the relative positions of pointer and normal happen to be. Repeat with the cell about 180° from its original position. The mean of the two values for $2\theta°$ should be taken if they are different. They probably will be different if the spindle is not exactly in the centre of the scale. The refractive index is $\frac{1}{\sin\theta}$.

III. *Spectrometer Method,* using hollow prism. See a later section (p. 166).

VIBRATION

A few connected experiments illustrating different types of vibration.

VIBRATION OF A STRETCHED STRING

THE frequency of vibration of a stretched string sounding its fundamental is given by the formula

$$n = \frac{1}{2l}\sqrt{\frac{T}{m}}$$

where n = frequency, or number of complete vibrations per second, l = length in cms., T = tension in dynes and m = mass in grms. per cm. length. The sonometer, illustrated in Fig. 42, is a stringed apparatus which has one movable

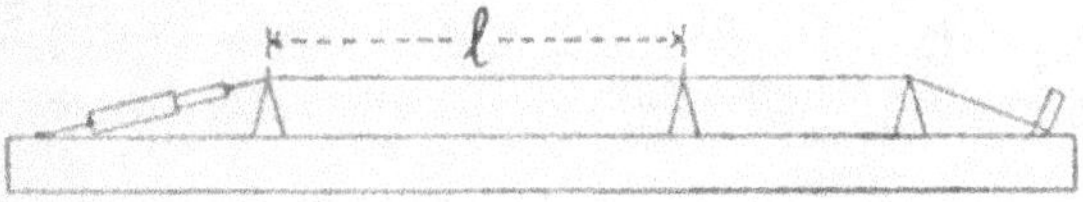

FIG. 42.

and two fixed bridges, a wrest pin and spring balance, the latter for indicating the tension (which must be converted to dynes). The string may be tuned to unison with a tuning fork by altering l and T as required.

To verify the above formula, show

(*a*) That $l \propto \frac{1}{n}$ (or nl = *constant*), when T and m are kept constant. Do this by tuning to unison with several forks whose values of n are known. Tune by altering l only, by means of the movable bridge. Tabulate nl for the several cases. The values should be identical.

(*b*) That $\sqrt{T} \propto n$, (or $\frac{\sqrt{T}}{n} = constant$) when l and m are kept constant. Here, alter T only, by means of the wrest pin. Keep l constant. Tabulate values of $\frac{\sqrt{T}}{n}$ for the different cases.

It could also be shown, by using different strings, that $n \propto \sqrt{\frac{T}{m}}$ when l is kept constant, by tuning each string, by means of the wrest pin only, to a given fork, and showing that $\sqrt{\frac{T}{m}}$ remained always the same whatever the value of m might be. Perhaps it is sufficient to know that this *could be done.*

Supposing, then, that we have verified the relation,

$$n \propto \frac{1}{l}\sqrt{\frac{T}{m}}$$

we can show that the constant 2 in the denominator of the original formula is correct in the following way. Take a single fork of known frequency n, and obtain several values of $\frac{\sqrt{T}}{l}$ (by tuning different lengths of string to this fork by means of the wrest pin), and take the mean of these values. Then weigh a length of the string, or one of the same kind, to get m. Put this in, together with the constant 2 and verify that $\frac{1}{2l}\sqrt{\frac{T}{m}}$ actually equals the known frequency of the fork. Assuming the truth of the formula for future use, we can then find the frequency of any fork by an experiment similar to this last one.

Vibration of a Column of Air—Resonance Method

Longitudinal stationary waves of air in a stopped pipe. A pipe of given length may sound its fundamental, its 1st harmonic, its 2nd harmonic, and so on. Or pipes of different lengths may all sound the same note if their lengths be such that this note corresponds to, say, the fundamental of one, the 1st harmonic of another, the 2nd harmonic of a third, and so on. The lengths are approximately $\frac{\lambda}{4}, \frac{3\lambda}{4}, \frac{5\lambda}{4}$, etc., where λ is the wave length in air for the particular note. Since $V = n\lambda$ or $n = \frac{V}{\lambda}$ where V denotes the velocity of sound in air, each of these pipes may be made to sound a note of frequency $\frac{V}{\lambda}$. If a fork of this frequency be sounded at the mouth of one of the pipes, the pipe will resound to the note. To determine the lengths, use a vertical glass tube arranged as in Fig. 43, so that water can be raised and lowered in it. The lengths at which resonance occurs can be determined by sounding the fork at the mouth of the tube and altering the level of the reservoir shown in the figure. The first length is not exactly $\frac{\lambda}{4}$, but is about $\frac{\lambda}{4} - \frac{3}{5}r$, where r is the radius of the tube. The next length is $\frac{\lambda}{2}$ greater than the first, and so on. Or,

$$l_1 = \frac{\lambda}{4} - \frac{3}{5}r,$$

$$l_2 = \frac{3\lambda}{4} - \frac{3}{5}r,$$

$$l_3 = \frac{5\lambda}{4} - \frac{3}{5}r, \text{ etc.}$$

Thus

$$l_2 - l_1 = \frac{\lambda}{2}, \qquad l_3 - l_2 = \frac{\lambda}{2}, \text{ etc.}$$

λ should be found from these, and the value $\frac{3}{5}r$ (the "end correction") checked; it is of course the difference between $\frac{\lambda}{4}$ and the length l_1.

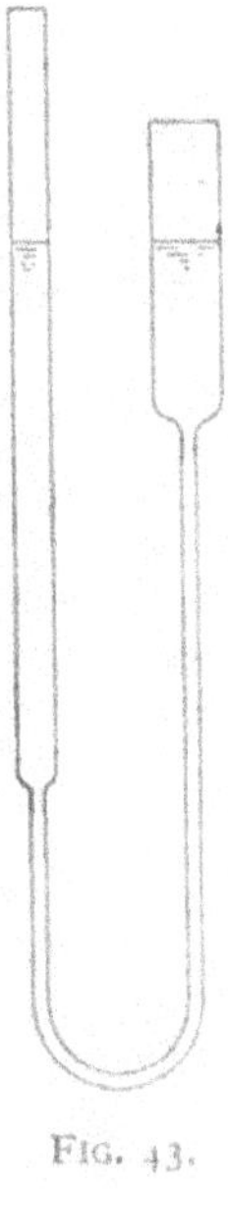

FIG. 43.

The experiment can be used to determine V if n be known, or n if V be known. If n be known, we have $V = n\lambda$ where V and λ are the velocity and wave length in air at the temperature of the experiment. The velocity at 0° C., say V_0, is equal to $V\sqrt{\frac{T_0}{T}}$ where T, T_0 denote the absolute temperature at which the experiment is performed, and the absolute temperature of 0° C., so that if the temperature of the experiment $= t°$ C. we have

$$V_0 = V\sqrt{\frac{273}{273+t}}.$$

The temperature of the air should be taken and V_0 determined. If n be the unknown to be determined, the value of V_0 must be looked up in the tables, V calculated by means of the above formula and divided by λ as obtained from the experiment. For very accurate determinations it would be necessary to allow for the fact that the air in the tube over the water surface is moist, a circumstance which has an influence on the velocity of sound.

The expression for velocity of sound in air from which the above equation for V_0 is derived is $V = \sqrt{\frac{\gamma P}{D}}$, P being pressure and D density of air, and γ a constant (the ratio of the specific heat of air at constant pressure to that at constant volume) whose value is 1.41.

Density of Air. The velocity formula may be verified by making a determination of the density and pressure of the air at the temperature at which the resonance experiment was made. A change in pressure between the two determinations is immaterial, as this does not alter V. It alters both P and D in the same ratio, and therefore the value of $\frac{P}{D}$ is unaltered. To make the determination, dry a bulb, Fig. 44, provided with a good tap, and let in atmospheric air. Weigh. Exhaust at the filter pump and weigh again. If exhaustion had been complete, the difference would have given the mass of air filling the bulb. But this is not the case. Open the tap under water, and hold the bulb so that the water rises to the same level inside as outside. The air inside is now at atmospheric pressure, and the water has replaced the air which was pumped out. Close the tap and weigh. Assuming,

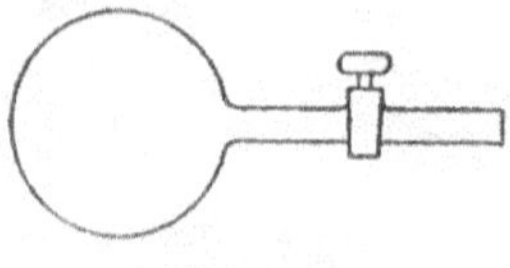

FIG. 44.

as we may for our purpose, that 1 cc. of water has mass of 1 grm., the mass of the water is numerically equal to the volume of the air pumped out. The mass of this air is known, and thus its density can be calculated. [The experiment is not quite accurate, on account of the presence of more water vapour in the air in the bulb after the water has been admitted. Mercury could be used instead if the bulb were not too large, but a fairly large bulb is necessary in order to give a reasonable mass of air. A small bulb might be used if the weighings were carried out with great accuracy. Or, by making the experiment and calculation somewhat more complicated, the effect due to the vapour could be allowed for.]

The pressure P must be obtained from the barometer and expressed in dynes per sq. cm. If $\mathbf{H}_0$ cms. be the corrected height, then $P = \mathbf{H}_0 \times 13.6 \times 981$ dynes per sq. cm. Fill in the formula for V and compare with that obtained by resonance at the temperature of the experiments.

There is, however, no need to make the density determination at the same temperature as the resonance experiment. Simply find D for any temperature and reduce to density at 0° C. : D_0 say, remembering that $\frac{D}{D_0} = \frac{T_0}{T}$. Then determine pressure as above, and fill in $V_0 = \sqrt{\frac{\gamma P}{D_0}}$, P being the actual pressure of the atmosphere when D was determined. It need not be the same as when the resonance experiment was done. Compare this V_0 with the V_0 obtained by the resonance experiment.

Longitudinal Vibration of a Rod

A rod clamped at the centre, and rubbed with a rosined or damped leather, vibrates like the air in an open pipe, with a node at the centre. The length of the rod is $= \frac{\lambda}{2}$, where λ is the wave length, in the material of the rod, of the note emitted. Set up and tune the sonometer to the rod. Determine n for the sonometer. Thus, alter the length of the string till it sounds in unison with a known fork of frequency say n_1. Then, both lengths of the string being known, n can be calculated. This is the frequency of the rod. If

V = velocity of sound in the rod, $V = n\lambda$ as before. V can thus be obtained.

The formula for V in a solid is

$$V = \sqrt{\frac{Y}{D}},$$

where Y is a constant, called Young's Modulus of Elasticity, for the material, and D the density. These constants may be found experimentally and the formula tested.

Young's Modulus. The vibrating rod suffers compression and extension. Young's Modulus, which is concerned in the velocity of sound in the rod, is the ratio of force per sq. cm. applied to elongate or shorten the rod to the strain $\left(\text{or } \frac{\text{change of length}}{\text{original length}}\right)$ produced. For wires, the modulus is easily determined by noting the elongation produced by a given weight. The same method is applied to rods in engineering laboratories. It can, however, be shown that Y is concerned in the bending of beams, for when a beam is bent as in Fig. 45 the upper portion is shortened and the lower portion lengthened. The amount of bending strain produced by the load must therefore depend on Y. If the rod be rectangular, of length l, breadth b and depth d, the deflection at the centre due to a load there, when the rod is supported on Λ-blocks at the ends, is

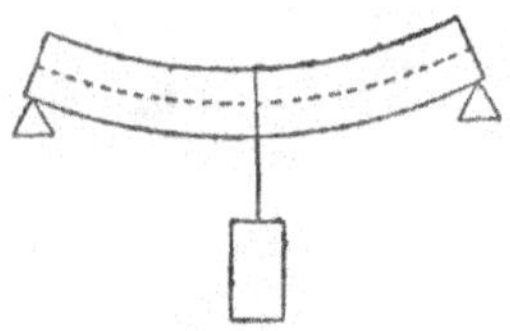

FIG. 45.

$$\delta = \frac{Wl^3}{4bd^3Y},$$

W being the load in dynes. For a rod of circular section, diameter d, the formula is

$$\delta = \frac{4Wl^3}{3\pi d^4 Y}.$$

Before determining Y, verify the facts:

(*a*) that $\delta \propto W$, l being constant, and

(*b*) that $\delta \propto l^3$, W being constant.

A convenient form of apparatus is sketched in Fig. 46. The rod is supported by stands at distance l apart. At the centre a thread is attached carrying a scale pan. Another thread (which must be quite distinct from the first) is also attached to the rod and passed completely round the small roller shown in the figure. A small weight keeps the thread taut. The thread may be rubbed with rosin to make it grip the roller. The roller works smoothly on a spindle to which a pointer is attached. This moves over a graduated circle. Different weights may be put in the pan, and the deflections noted. Remove the weights carefully and see whether the pointer returns to its original position. If not, the thread has slipped, or the rod has been too much strained—beyond its elastic limit. The deflection of the rod is related to the deflection of the pointer, for a complete rotation of the latter corresponds to a dip of the rod equal to one turn of the thread on the roller, i.e., to a dip in cms. of $2\pi r$, where r = radius of

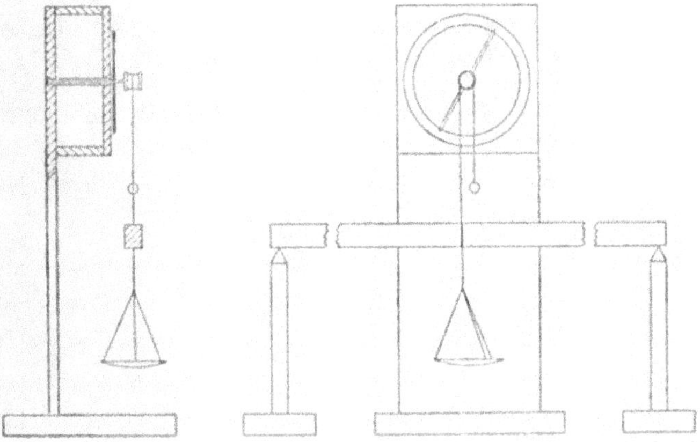

Fig. 46.

roller. This radius must be determined by means of a screw gauge or calipers, as must also the lateral dimensions of the rod.

The deflections should be taken for several lengths (i.e., distances between the supports of the rod) and for several weights for each length. Graphs should be plotted, on the same diagram, for each length. Ordinates will represent deflections of the rod, and abscissæ weights. Again, for each weight a graph should be drawn with lengths for abscissæ and deflections for ordinates, and also one with values of l^3 for abscissæ and deflections for ordinates. This last graph will be a straight line if $\delta \propto l^3$.

Next calculate $Y\left(\text{i.e., } \frac{Wl^3}{4bd^3\delta}\right)$, making a separate calculation for each length used. Thus, for the first length find the mean value of $\frac{W}{\delta}$ from the observations, and calculate the corresponding value of Y. Similarly for the other lengths. Compare these values of Y, which should be nearly alike.

Density of the Rod. Take the rod, or a sample of the material of which it is made. If of a regular shape, measure with vernier calipers or screw gauge and calculate the volume. Determine the mass, and thence the density. Also, if a sample is available, weigh in air and in water, and calculate the density, remembering that the difference between the weighings gives the mass of the water displaced. The first method will be sufficient if the material be wood.

Apply the values obtained for Y and D to the velocity formula, and compare with the velocity already determined by experiment.

EXPANSION

I. Solids

LET L, L_0 be the lengths of a rod at $t°$ C. and 0° C. Then $\frac{L - L_0}{L_0 t}$ is very nearly constant, and is called the coefficient of linear expansion of the solid of which the rod is made. Let it be denoted by l. We may write

$$L = L_0 (1 + lt).$$

The coefficient of expansion should be determined for the rods (omitting the wooden ones) used in the experiments on the vibration of rods and Young's modulus. A convenient apparatus for the purpose is shown in Fig. 47. The

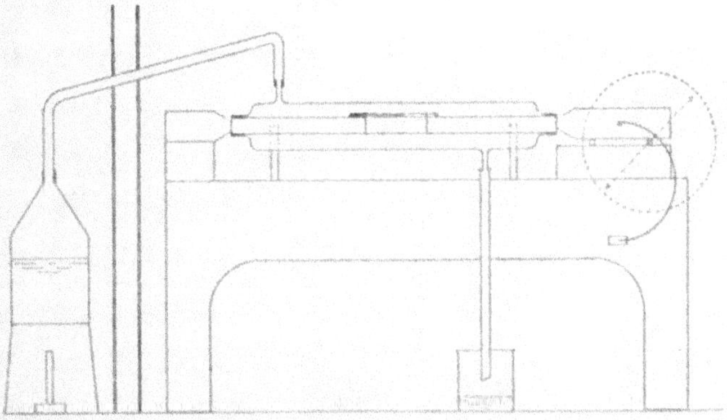

Fig. 47.

rod to be used is corked into a glass steam jacket, a thermometer being attached to the rod in such a way that it can be read from outside. The whole is supported by forks on a framework, between two jaws which touch the ends of the rod. One of these is fixed, the other moves on two small rollers over a flat plate, and is held in position by two semicircular springs, one of which is shown in the figure. These springs cause the jaw to press rather tightly on the rod and also on the rollers. The latter consist of thin needles or steel wires, and one of them is rather long and carries a light pointer, which moves over a graduated circle, shown dotted, as the jaw recedes owing to expansion of the rod. The expansion can be deduced from a knowledge of the diameter of the roller and of the angle through which it turns. Note that for one complete revolution of the roller the jaw moves horizontally a distance equal to *twice* the circumference of the roller. For the roller moves along the plate on which it rests through a distance equal to its own circumference, while the jaw also moves a similar distance relatively to the roller.

To determine l, note the temperature of the rod after the apparatus has been set up and left for a few minutes to assume a steady temperature. Note also the readings of both ends of the pointer. Pass steam for several minutes, till the pointer becomes quite steady, showing that expansion has ceased. If the thermometer can be read, observe its temperature; if not, read the barometer and determine from the boiling-point tables the boiling point of water, and assume that this is the temperature of the rod.

Let t_1 t_2 be the initial and final temperatures of the rod, and L_1 L_2 its lengths at these temperatures. Then from the general formula we have

$$L_1 = L_0 (1 + lt_1)$$
$$L_2 = L_0 (1 + lt_2).$$

The increase of length due to heating from $t_1°$ to $t_2°$ is $L_2 - L_1$ in cms., and we may write

$$L_2 - L_1 = L_0 (1 + lt_2) - L_0 (1 + lt_1)$$
$$= L_0 \, l \, (t_2 - t_1)$$

whence

$$l = \frac{L_2 - L_1}{L_0 (t_2 - t_1)}.$$

Since L_0 is not known, the experiment does not enable us to determine l. But if the original length L_1 (measured beforehand by a meter scale) be substituted for L_0 in the denominator, only an extremely minute difference will be made in the value of the fraction. This value may be taken for l for all practical purposes.

Care must always be taken in making approximations of this kind, otherwise the results obtained may be quite inadmissible. Suppose in the above formula $L_0 = 100.0$, $L_1 = 100.1$, and $L_2 = 100.2$ cms., these figures being taken at random for explanatory purposes only. Now $L_2 - L_1 = 0.1$ cm. and $\frac{L_2 - L_1}{L_0} = \frac{0.1}{100.0} = 0.001$. If in the denominator we substitute L_1 for L_0 we have $\frac{0.1}{100.1}$, which $= 0.000999$, which only differs from the true value by one part in a thousand. But now, supposing that we substituted L_0 for L_1 in the numerator, we should have

$$\frac{L_2 - L_0}{L_0} = \frac{100.2 - 100.0}{100.0} = \frac{0.2}{100.0} = .002,$$

which is twice the correct value. In the actual case, of course, the value $L_2 - L_1$ is determined directly by experiment, and so there is no danger of falling into this error.

At the end of the experiment allow the apparatus to cool, if possible, to the original temperature, and note whether the pointer comes back to its original position. If not, the experiment cannot be relied on as having given a satisfactory result.

A calculation

Consider one of the rods for which Young's Modulus and the coefficient of linear expansion have been measured. Suppose it to be heated to $t°$ C., and then let its ends be firmly fixed so that it is prevented from contracting as it cools. If the rod be cooled to 0° C., what force will it exert on the supports? or, what comes to the same thing, what will be the tension in the rod? This question is merely introduced here as a type of problem which connects two physical quantities or properties, and helps to show the relation between them.

II. Liquids.

A. *Coefficient of Absolute Expansion of Mercury by Dulong and Petit's Method.* The method is one in which two columns of mercury at different temperatures are balanced in a kind of U-tube. A simple arrangement is shown in Fig. 48. The two bent tubes, the vertical parts of which are surrounded by water baths, are connected together at the bottom by a length of capillary tube, which may be water-cooled in order to prevent flow of heat from the hotter to the cooler side of the U. Stirrers are provided for the baths, also thermometers should be suspended in them. The capillary tube being horizontal, and the baths kept at steady temperatures, say $t_1°$ and $t_2°$, the difference between the levels of the mercury surfaces should be accurately determined by means of a reading microscope. The lengths of the portions of the tubes in the baths, which should be as nearly as possible equal, will also be required. 60 cms. is a suitable length. A steam jacket may be used with advantage in place of the water bath for the high temperature side of the apparatus.

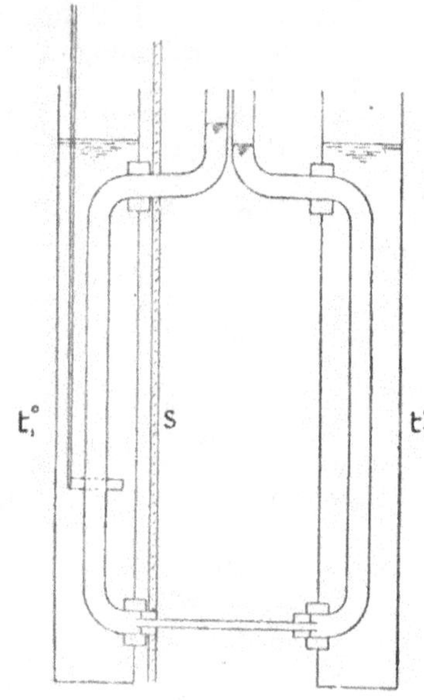

FIG. 48.

Let h be the level difference and H the vertical length of the portion of either tube in its bath, and let ρ_1, ρ_2 be the densities of mercury at $t_1°$ and $t_2°$, and suppose for simplicity $t_2°$ to be the temperature of the air and of the parts of the tubes to the right of the screen S in the figure. The pressure due to the column of mercury in the hot part of the left hand tube is $g \rho_1 H$, and that due to the corresponding column on the right side is $g \rho_2 H$. This is greater, but the difference is made up by the excess column, of height h, at the mouth of the left tube. The pressure due to this is $g \rho_2 h$. (The total pressure on the left side must equal that on the right side, since the tubes are connected by the horizontal capillary tube at the bottom.) Thus we have

$$g \rho_2 H - g \rho_1 H = g \rho_2 h$$

or

$$(H - h) \rho_2 = H \rho_1$$

whence

$$\frac{H - h}{H} = \frac{\rho_1}{\rho_2}.$$

It is only necessary to express the ratio of the densities in terms of the coefficient of volume expansion of mercury. Taking say 1 grm. of mercury at different temperatures, its density will vary inversely as its volume, or $\frac{\rho_1}{\rho_2} = \frac{v_2}{v_1}$ where v_1, v_2 represent its volume at $t_1°$ and $t_2°$ respectively. But if m denotes the coefficient of expansion and v_0 the volume at 0° C. we have

$$v_1 = v_0 (1 + mt_1) \qquad v_2 = v_0 (1 + mt_2),$$

and therefore we easily arrive at the required result

$$\frac{\rho_1}{\rho_2} = \frac{1 + mt_2}{1 + mt_1},$$

and finally

$$\frac{H - h}{H} = \frac{1 + mt_2}{1 + mt_1},$$

from which m can be calculated.

B. *Coefficient of Apparent Expansion of Mercury in Glass. Weight Thermometer Method.* Let m_a denote the coefficient of apparent expansion of mercury in a vessel of glass, and g the coefficient of cubical expansion of glass. Then

$$m = m_a + g.$$

If some other liquid be considered whose coefficients of absolute and apparent expansion (the latter in respect to a vessel of the same kind of glass as before, or preferably to the same vessel) are c and c_a we have

$$c = c_a + g.$$

Now the determination of c_a is simpler than that of c, and consequently it will be convenient to determine m and m_a, and to calculate g from these, and then to determine c_a for any liquids whose c may be required. m being already known, it remains to determine m_a for mercury in a given vessel.

Use a glass bulb of the shape shown in Fig. 49, say about 3 inches long. Weigh it dry and empty, then fill with mercury by alternately warming and cooling with the nozzle under mercury. Immerse the bulb in a beaker of pounded ice and distilled water, in the position shown in the figure, and when all is at 0° C. remove the mercury cistern, and replace by a small weighed empty beaker. The quantity of mercury present is then just sufficient to fill the "weight thermometer" at 0° C. Next immerse in a bath at a higher temperature $t°$ (e.g., 50° C.). Keep the mercury which overflows, remove this in its small beaker, and then allow the weight thermometer to cool. Weigh both it and the overflow, deducting the known weights of the bulb itself and of the beaker. Let W_t, W_0 denote the masses of mercury in thermometer at $t°$ and 0° respectively. The latter of course is equal to the mass left in at $t°$ plus the overflow.

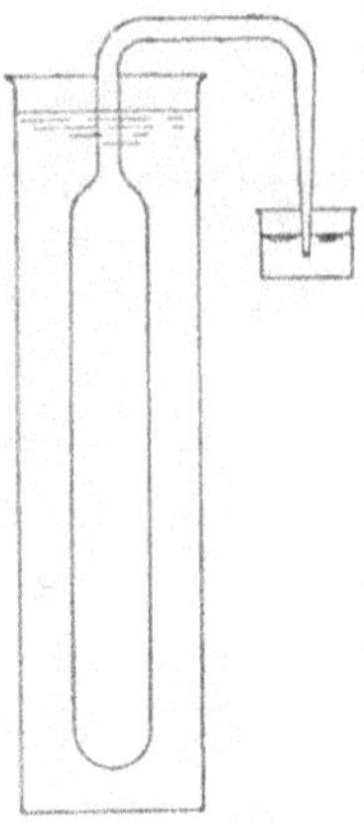

FIG. 49.

Now the coefficient required, m_a, is given by

$$m_a = \frac{\text{mass of overflow}}{\text{mass filling bulb at } t° \times \text{rise of temperature}}$$

or

$$m_a = \frac{W_0 - W_t}{W_t \times (t° - 0°)}.$$

Nearly the same result will be obtained if the temperature of the room be taken instead of 0° C. The value of g can now at once be found, and the bulb used for other liquids.

C. *The same for any liquid.* The next step is to determine the value of c for any liquid, using the same bulb and method as before.

D. *Calculation of Coefficient of Absolute Expansion for the Liquid.* Since g and c_a are known, c can at once be determined. To show how the various determinations that have been made assist in the final result we may write:

$$c = c_a + (m - m_a).$$

Other liquids may then be substituted, and their c_a's determined—the same value of $m - m_a$ (or g) of course serving for each case.

As an exercise on the subject of this section the coefficient of linear expansion of a glass tube may be determined, and multiplied by three, to give the cubical expansion g. A weight thermometer may then be made from this tube, and g determined by means of mercury. The two values thus obtained should then be compared.

EXPERIMENTS ON WATER SUBSTANCE

Change of State I. (Ice-Water)

Determination of the Latent Heat of Water. The latent heat of water (or of fusion of ice as it is sometimes called) is given by the number of calories absorbed by one gram of ice in becoming water without change of temperature. Use a calorimetry method, and first perform a preliminary experiment in the following way: Half fill a calorimeter with warm water, and note its temperature. Put some small lumps of ice into the water, and stir till these are melted. Note the reduction of temperature. Estimate roughly from this the quantity of ice necessary to cause a reduction of temperature of about 15° C.

Now perform the actual experiment for the determination of latent heat. Weigh an empty calorimeter, and the saime half full of water, which should be at a temperature of 7° or 8° above that of the room. Suspend the calorimeter in its outer cover. Take the temperature accurately, and rapidly transfer to the water about the quantity of ice estimated as above. This ice should be carefully dried on blotting paper immediately before being put into the calorimeter, otherwise any adhering water will give rise to error. Stir with the thermometer till the ice is melted, and note the final temperature. This should be about as much below the atmospheric temperature as the initial temperature was above it. The reason for this will be explained shortly. Now equate the heat lost by the calorimeter and its original quantity of water to that gained by the ice in melting and by the melted ice in rising to the final temperature of the calorimeter. If E denotes the water equivalent of the calorimeter, that is, its mass $\times$ specific heat, and M the mass of water originally present, the heat lost will be $(M + E) \times$ (fall of temperature), and if $m =$ mass of ice added (which can be determined by making a final weighing of the calorimeter and its contents, and deducting from this the mass before the ice was put in), and $L =$ the latent heat, the heat gained will be equal to $mL + m \times$ (rise of temperature from 0° to final temperature of the water). The object of arranging the original temperature and the quantity of ice so that the final temperature is approximately as much below atmospheric temperature as the initial temperature was above it, is to minimize error due to loss of heat from the calorimeter during mixing. Since during half this process the calorimeter is above the temperature of the room, and is below it during the other half, the slight loss during the first half will be approximately made up by a gain during the second half.

Change of State II. (Water-Steam)

Determination of the Latent Heat of Steam at Atmospheric Pressure. The latent heat of steam depends on the temperature at which boiling takes place, being smaller the higher the temperature. The temperature of boiling (in a vessel open to the air) may be obtained by observation of the height of the barometer and reference to the tables of boiling points. The latent heat is given by the number of calories absorbed by one gram of water in passing to steam without change of temperature. The same quantity of heat is given out when the steam condenses. The method used is to pass steam into a calorimeter of water, where condensation takes place. The loss of heat during the condensation must be added to the loss which the condensed water suffers in being cooled to the final temperature of the calorimeter. The loss is then equated to the heat gained by the calorimeter and the water originally contained in it. The mass of steam passed in is determined by weighing before and after the process.

One or two points call for special attention. In the first place, the final temperature must not be so high as to give rise to an appreciable loss of water by evaporation from the calorimeter, an occurrence which would obviously vitiate the results of the experiment. A considerable rise, however, is advisable, so that the quantity of steam condensed may be sufficiently great to enable its mass to be obtained with accuracy. If the calorimeter be cooled to about 10° below atmospheric temperature, by means of ice, before passing steam, and this be passed till a temperature of 10° above atmospheric is reached, a total rise of 20° is obtained safely, and at the same time the advantage referred to in the last experiment is also gained. Another point is that the steam which is passed into the calorimeter should be as dry as possible, that is, it should not be already partly condensed to water before entering the calorimeter. This will certainly occur unless special precautions be taken to prevent it. A suitable arrangement is shown in Fig. 50. An inverted flask, in which the steam is to be generated, is fitted with a cork, pushed into the neck as shown, and a glass delivery tube. The upper part of this tube being at the temperature of boiling water, the steam which passes out is largely kept from condensing, provided that the lower part of the tube is

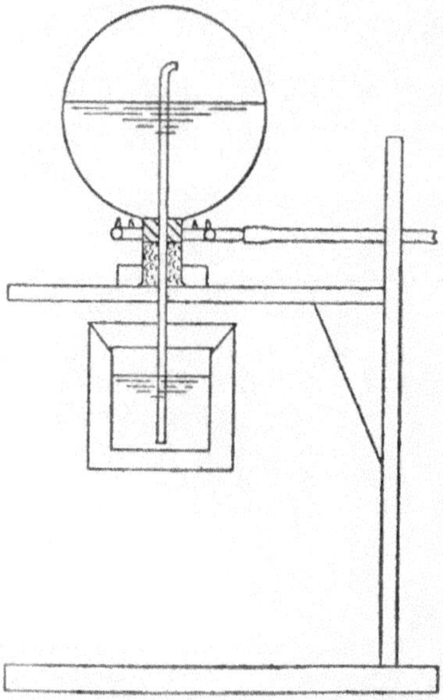

FIG. 50.

not longer than is necessary to enable it to dip to the bottom of the calorimeter. The flask is supported on a piece of wood which acts as a screen to prevent heat from reaching the calorimeter direct. The steam is generated by means of a ring burner. The neck of the flask should be rather short and the part below the cork stuffed with cotton wool. The reason for pushing in the cork to the end of the neck is that water in the neck would be rather below boiling point, and would therefore tend to make steam in the outlet pipe condense. In performing the experiment, the calorimeter of water must be brought quickly into position, and quickly removed when sufficient steam has been passed. (If the apparatus be such that the thermometer cannot be kept in the calorimeter so as to show when the passage of steam is to be stopped, a preliminary experiment might be made in order to find out how many seconds should be allowed so as to obtain about the required rise of temperature.) Let m be the mass of steam condensed, t_1 its original temperature, t_2 the initial temperature of the calorimeter, and t_3 the final temperature of the whole. Then L being the latent heat of steam, we have

$$mL + m\,(t_1 - t_3)$$

for the heat given up, and

$$(\text{mass of water originally in calorimeter} + E) \times (t_3 - t_2)$$

for the heat gained, where E as before denotes the water equivalent of the calorimeter. Equating these, L can be determined.

Vapour Pressure

The steam dealt with in the last experiment is saturated water vapour whose pressure is equal to that of the atmosphere, and whose temperature is that of the boiling point of water at atmospheric pressure. It represents, then, saturated water vapour at a particular temperature and pressure. By causing water to boil under various pressures and noting the corresponding boiling points, data for drawing a graph or constructing a table showing the relation between pressure and temperature of saturated water vapour can be obtained. Such a table has already been referred to above. A convenient apparatus for pressures lower than atmospheric is shown in Fig. 51. Water is boiled in a flask provided with a thermometer, whose bulb is a little above the surface of the water, and an outlet tube surrounded by a condenser, through which cold water circulates. This condenses the steam, which drops back as water to the flask. The outlet tube passes into a bottle which also has another tube fitted through its stopper (which should be of rubber, as should also that of the flask, for the sake of air-tightness) passing to a mercury pressure gauge and on to a filter pump. By means of the pump the pressure in the flask can be reduced below that of the atmosphere. The bottle serves two purposes: it prevents any condensed steam which may pass over from reaching the pressure gauge, and it also enlarges the volume of the exhausted system. This is convenient in practice, as it causes changes of pressure to take place more slowly.

Suppose the water to be boiling and the pump to be working. By means of a tap between the apparatus and the pump the pressure can be adjusted to the required value

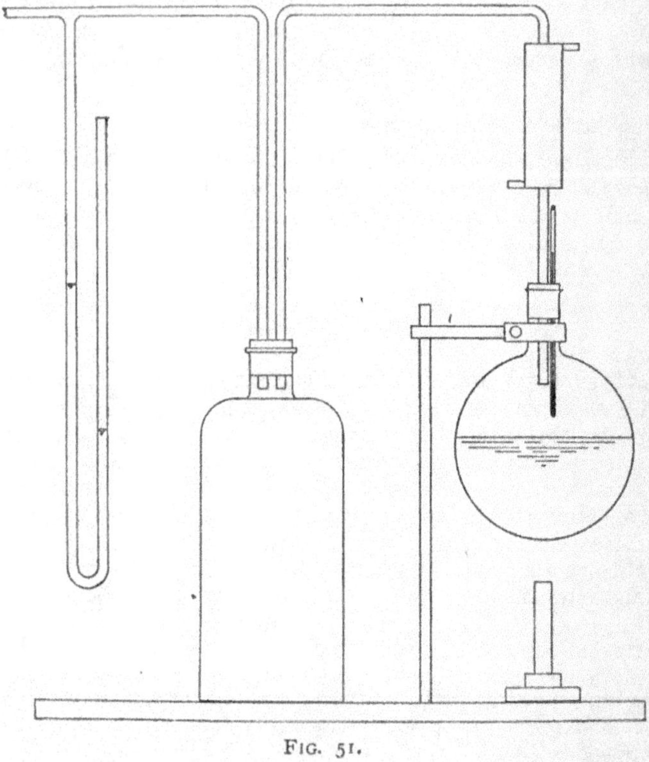

FIG. 51.

and kept steadily at that value, as indicated by the gauge, till the temperature is read on the thermometer. The pressure and temperature will then give a point on the graph previously referred to. A number of values, using the greatest range of pressures possible, should be observed and plotted. The results should then be compared with the standard boiling point or vapour pressure tables. Note that the pressure in the flask is given in cms. of mercury by the height of the barometer at the time of the experiment minus the difference between the levels of the mercury in the gauge. One determination should be made without working the pump, so as to include the case of boiling under atmospheric pressure.

Another method may be used for temperatures ranging from just above freezing point up to 60° or 70°. The necessary apparatus is shown in Fig. 52. It consists of two barometer tubes, originally filled with mercury and inverted into the same cistern. Into one of them is passed a small quantity of water, which rises to the top of the mercury column and partially evaporates, filling the upper part of the tube with saturated water vapour. The upper parts of the tubes are enclosed in a water bath with a glass front. A scale dipping into this bath, a stirrer, and a thermometer to indicate the temperature of the bath are also required. Water at various temperatures may be poured into the bath. Were it not for the presence of the water and its vapour in one of the tubes the two mercury columns would stand at the same level. The depression of the mercury in the water tube is due partly to the small quantity of water, which exerts a pressure (expressed in cms. of mercury) equal to its height divided by 13.6—the pressure due to a column of water being equal to that due to a column of mercury $\frac{1}{13 \cdot 6}$ times its height. This only accounts for a very small depression of the mercury, the greater part being due to the pressure of the water vapour in the upper part of the tube. This pressure, then, can easily be determined by adding to the mercury column in the tube an amount equal to $\frac{1}{13 \cdot 6}$ of the height of the small column of water, and then subtracting this total height from the height of the column in the other tube. Since only differences of the heights of the columns are required, it is immaterial where the zero of the scale used for measuring them may be. The result is the pressure of the vapour in terms of cms. of mercury, the mercury being at the temperature of the bath. To reduce to standard cms. of mercury (in which the mercury is supposed to be at 0° C.) it is necessary

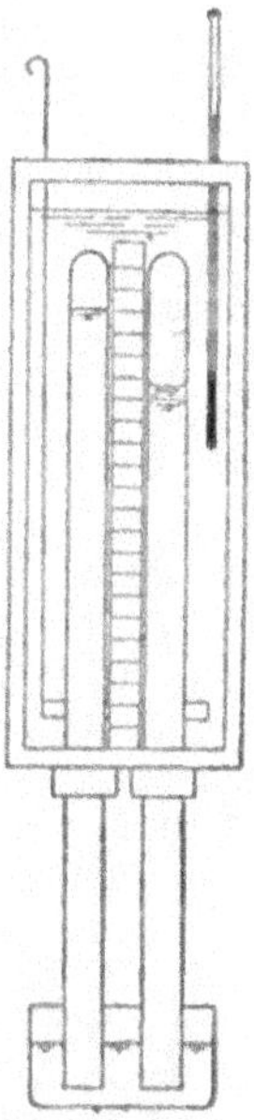

FIG. 52.

to divide the value as determined above by $1 + mt$, where m is the coefficient of expansion of mercury and t the temperature of the bath. A set of values of pressure and temperature can thus be obtained and plotted on the diagram of the last experiment.

WATER VAPOUR IN THE ATMOSPHERE—HYGROMETRY.

Dew Point. In the last experiments a curve showing the pressure of saturated water vapour at different temperatures has been obtained. In the atmosphere, part of the ordinary atmospheric pressure is due to the water vapour present. If at any temperature the quantity present be such as to exert a pressure equal to the saturation pressure (as shown on the curve) for that temperature, the humidity of the atmosphere is the greatest possible. Any excess of moisture would condense in the form of dew. If the air contains less moisture than this, the "relative humidity" is equal to the ratio of the pressure of the vapour present to the pressure which would be exerted if the vapour were saturated, or, in other words, is equal to the ratio of actual pressure to maximum pressure possible. This maximum pressure, for the particular temperature of the atmosphere, is given by the graph or by the tables of saturated vapour pressure. To find the actual pressure, imagine the temperature of the atmosphere to be reduced (without altering the amount of vapour present) till condensation just begins to take place. Then the pressure actually present is the maximum pressure for the lowered temperature, and its value can be found from the graph or tables. The ratio of these two pressures gives the relative humidity of the atmosphere at the original temperature. The lower temperature is called the Dew Point. It obviously depends on the quantity of vapour present in the atmosphere.

To determine it practically, use Regnault's hygrometer, which consists of a thin polished silver thimble cemented on to the end of a glass tube, fitted with a cork, a thermometer, with its bulb in the middle of the thimble, and two tubes, one passing down nearly to the bottom of the thimble, and the other, an outlet tube, passing just through the cork. (See Fig. 53.) To use the apparatus, pour ether into the thimble, and pass a stream of air gently down the inlet tube by means of a small hand bellows, or draw through the outlet tube by an aspirator. Air bubbles through the ether, and induces evaporation, and so causes a lowering of the temperature of the thimble and its contents. When the dew point is reached a film of dew is seen to form on the outside of the thimble by condensation from the atmosphere. Stop the stream, and note the temperature indicated by the thermometer. Also note the temperature at which the dew disappears. Any appreciable difference between these temperatures indicates that the cooling has been too rapid (in which case the thermometer may not take up quickly enough the exact temperature of the ether and thimble), or else that the first trace of dew has not been observed. It is not, in fact, easy to detect this. Sometimes a second (dummy) thimble is placed near the first for comparison, and both are observed through a telescope. In any case it is necessary that no moisture due to the breath of the observer should reach the apparatus. A large glass screen may be used to prevent this. If no telescope be used, the film of moisture will be more readily detected if the thimble be continually stroked with a paper spill—the track of this will be visible as soon as the film forms.

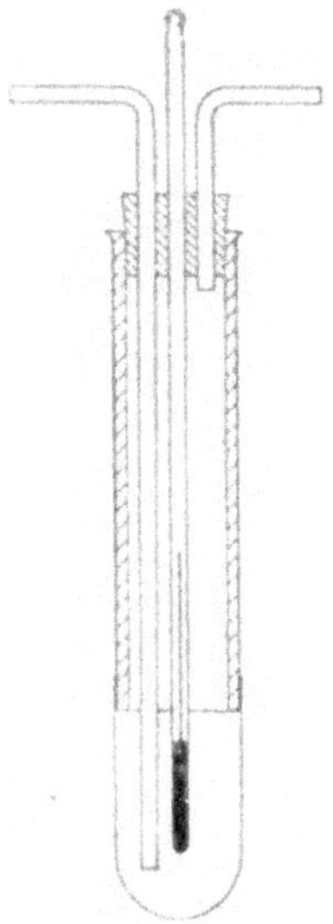

FIG. 53.

Wet and Dry Bulb Hygrometer. Another form of instrument for determining relative humidity is the wet and dry bulb hygrometer. Two thermometers are supported side by side on a stand, the bulb of one of them being kept moist by means of a cotton wick, one end of which dips into a small vessel of water—the vapour from which should not be allowed to reach the bulbs, or else the hygrometer will register too high a value for the humidity of the atmosphere. The wet thermometer generally shows a lower temperature than the other, due to evaporation of water from its bulb, which causes a reduction of temperature. If, however, the air is already quite saturated, there will be no evaporation and therefore no reduction of the temperature of this thermometer. Equality of temperatures thus means that the air is saturated. For other cases, tables have been prepared which show the pressure of vapour present for all readings of the two thermometers. These have been prepared by noting the readings of the thermometers for different states of humidity as determined by means of other hygrometers, such as the one previously described. Thus the wet and dry bulb hygrometer is not an instrument from which the humidity can be directly obtained, but depends for its value on results obtained by other means. It is, however, far simpler for practical use than Regnault's form, once the necessary tables have been compiled.

RADIATION

THE rate at which a body loses heat by radiation depends on the nature and extent of the surface, and on the temperature of the surface and that of the enclosure in which it is placed. Simple experiments can be performed on cooling by means of tin vessels coated with different materials, containing hot water constantly stirred. These may be suspended from retort rings, as shown in Fig. 54, by strings attached to little hooks soldered

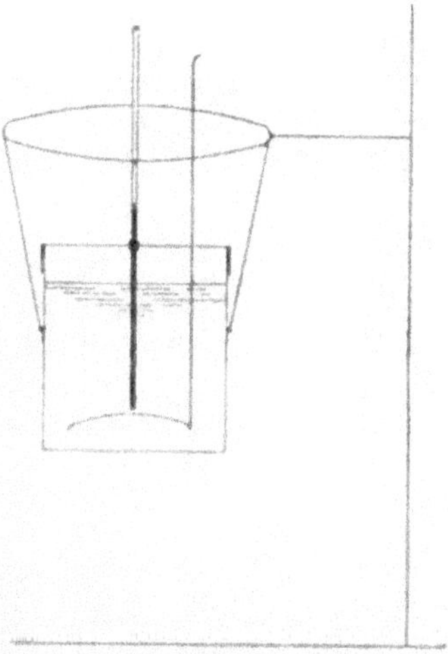

FIG. 54.

to the tins. They should be protected from draughts, and the room temperature may be taken as that of the enclosure. Of course, some heat is carried away by connexion currents in the air; nevertheless the experiments are instructive. Cooling curves, temperatures as ordinates and times as abscissæ should be drawn, using the different vessels successively. These vessels should be all alike except as regards their surfaces. One may be bright, one covered with soot over a smoky flame, another painted white, and so on. The varying Emissive Powers will be indicated by the different rates of cooling obtained.

Newton's law of cooling states that the rate for a given surface is proportional to the difference between the temperature of the surface and that of the enclosure. This law is approximately true so long as this difference is not great. The law should be tested for each surface, or at least for those of greatest and least emissive powers. Having a graph of cooling, we may find the rate of cooling at any instant by drawing a tangent to the curve at the point which represents that instant. For example, to find the rate of cooling at a

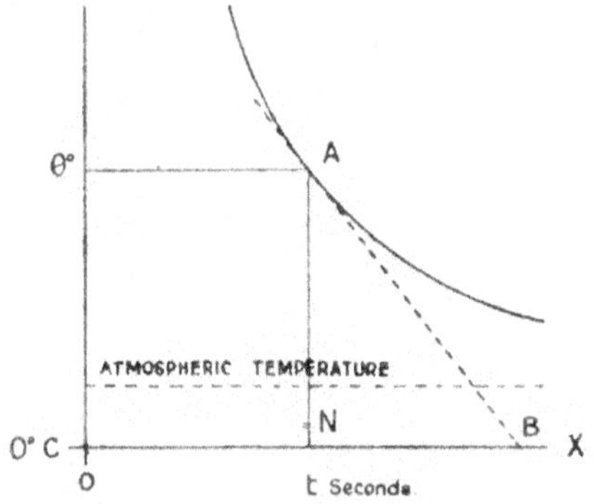

FIG. 55.

time t from the beginning of the observation, a tangent must be drawn at the point A, Fig. 55, corresponding to t on the "time-axis" O X. If the cooling continued at the same rate as at A, the cooling curve would be represented by the tangent A B. As a matter of fact, the cooling slows off as time goes on and as the body gets nearer and nearer to atmospheric temperature. The graph curves in such a way as to become finally a horizontal straight line corresponding to atmospheric temperature, when cooling ceases. The rate of cooling at t then is represented by the straight line A B ; thus, A N°* or $\theta°$ would be lost in time corresponding to the distance N B on the time-axis. If A N represents 80° C. and N B 10

* The point N is where the vertical through A cuts O X.

minutes or 600 seconds, the body is cooling at the rate of 80° C. in 600 seconds, or $\frac{80}{600}$ degrees in 1 second. This is the rate of cooling at A, i.e., at time t. It is in fact $\frac{AN}{NB}$, the numerator and denominator being taken to their proper scales, and not merely measured in cms. on the graph.

Now, Newton's law says that this rate is proportional to the excess of temperature of the body over that of the enclosure, say the room. To test this draw tangents at several points on the curve. Calculate rate of cooling for each, and also read off the temperature at the point and deduct temperature of room, which will give the excess required. Let C_1, C_2, C_3 be the rates of cooling, and $\theta_1-\theta$, $\theta_2-\theta$, $\theta_3-\theta$ be the differences between the temperature and that (θ) of the room. Then the law states that

$$C_1 : C_2 : C_3 = \theta_1-\theta : \theta_2-\theta : \theta_3-\theta$$

or

$$\frac{C_1}{\theta_1-\theta} = \frac{C_2}{\theta_2-\theta} = \frac{C_3}{\theta_3-\theta}.$$

These fractions may be calculated, and their degree of approximation to equality tested.

The radiation, the amount of which depends on the factors we have been considering, travels off in straight lines, and its intensity in space, after leaving its source, decreases. The intensity of radiation from a point source decreases according to the inverse square law. If the body be an extended one, each little bit of surface may be supposed to emit radiation, which follows this law.

Inverse Square Law

I. *Source, hot metal ball.* An iron or copper ball (Fig. 56) hung by a chain in a Bunsen flame forms a convenient source of radiation. If the temperature of the ball were uniform all over its surface it would emit symmetrically in all directions and would be equivalent for our present purpose to a point source situated at its centre and emitting the same amount of radiation. If it be not uniformly heated, it will still, for points a considerable distance away, behave nearly like a point source at its centre. To measure the intensity of the radiation at any distance use a thermopile and galvanometer.

The deflections of the galvanometer are nearly proportional to the intensity of radiation falling on the pile. If the pile be provided with a funnel, polished inside, so that approximately all the radiation which falls on the mouth of the funnel reaches the pile, the distance from the centre of the ball (i.e., from the equivalent radiating point) to the mouth

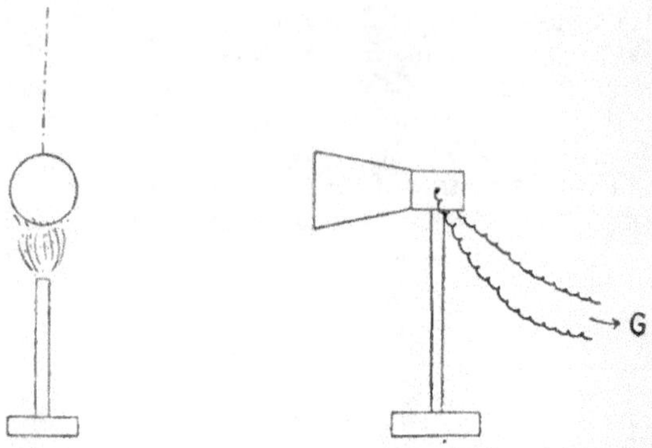

FIG. 56.

of the funnel must be taken. If no reflecting funnel be used the distance must be measured to the face of the pile. Setting the pile at various distances and taking deflections the law can be verified. We have, if $r_1\, r_2\, r_3$ be the distances, and θ_1, θ_2, θ_3 the deflections,

$$\theta_1 : \theta_2 : \theta_3 = \frac{1}{r_1^2} : \frac{1}{r_2^2} : \frac{1}{r_3^2}.$$

or

$$\theta_1 r_1^2 = \theta_2 r_2^2 = \theta_3 r_3^2.$$

The constancy of these should be verified.

Assuming the deflections to be proportional to the intensities, I_1, I_2, I_3 say, we may write

$$I_1 r_1^2 = I_2 r_2^2 = I_3 r_3^2.$$

II. *Source of Light.* The inverse square law can be verified in the case of light by means of a photometer—Bunsen's grease spot photometer is suitable. A plan of the arrangement is shown in Fig. 57. A piece of white paper having a spot of grease in the middle is supported in a small frame at A, placed on a long graduated rod. Blocks holding candles slide along this rod on each side. Mirrors M M are placed so that the eye at E can see images of both sides of the spotted paper. This arrangement must be symmetrical so that both sides are viewed from the same angle. Now, suppose both sides to be equally illuminated, evidently they will both appear exactly similar to the eye, provided the greased paper is the same on both sides, as it should be. But if one side be more strongly illuminated than the other, the plain paper will on that side look brighter than the grease spot; while the spot will look brighter than the paper on the

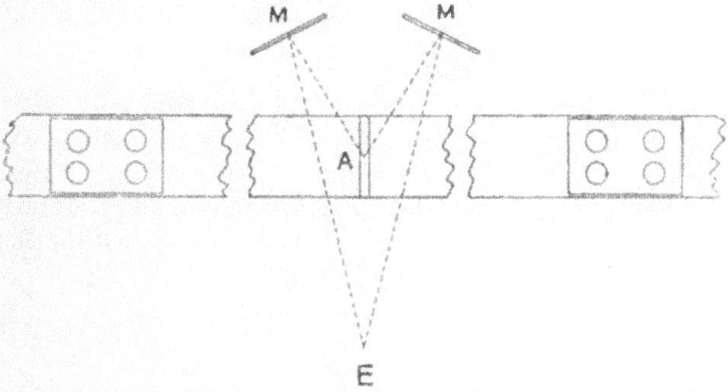

FIG. 57.

other side, due to the greater transparency of the greased portion. The arrangement then acts as a sort of illumination balance. Now, to verify the inverse square law, light one candle on one side and two on the other. Let the illuminating power of each candle be denoted by C. They should burn as evenly as possible. The intensity of illumination at a distance r will be given by

$$I = \frac{C}{r^2}$$

if the law be true for this luminous radiation. Adjust the photometer by moving the candle holders till the illumination is equal on the two sides. Let R_1, r_2, r_3 be the distances of the candles from the grease spot, then since the illumination is the same on both sides we have

$$\frac{C}{R_1^2} = \frac{C}{r_2^2} + \frac{C}{r_3^2} \quad \text{or} \quad \frac{1}{R_1^2} = \frac{1}{r_2^2} + \frac{1}{r_3^2},$$

for the illumination due to the two candles (2 and 3) is obtained by adding their contributions together. To determine the required distances, the positions on the rod of the edges of the sliding blocks may be noted and allowance made for the distances of the candles from these edges; the position of the central stand on the scale must also be observed. The mean distance of candles 2 and 3 may be used instead of the separate distances (provided that candles 2 and 3 are near together), and in this case, calling this distance R_2, we have

$$\frac{C}{R_1^2} = \frac{2C}{R_2^2} \quad \text{or} \quad \frac{1}{R_1^2} = \frac{2}{R_2^2}.$$

Repeat with three and with four candles instead of with two. Then

$$\frac{1}{R_1^2} = \frac{2}{R_2^2} = \frac{3}{R_3^2} = \frac{4}{R_4^2},$$

these fractions being proportional to the four illuminations, which have all been made equal by adjustment. Measure and square the R's, tabulate and see to what approximation they follow the proportion 1 : 2 : 3 : 4. Repeat using candle 1 at a different distance from the spot.

Illuminating Power—Candle Power

Having verified the inverse square law by the use of several equal sources of light, we may use it to determine the illuminating power or candle power of some other source, by comparison with a candle. Using the same arrangement as before, we may substitute the new source, say a gas flame, for the candles on one side, retaining one candle on the other. When equality of illumination is attained, we have, letting C^1 stand for the illuminating power of the new source and R_2 its distance,

$$\frac{C}{R_1^2} = \frac{C^1}{R_2^2}, \quad \text{or} \quad C^1 = \frac{R_2^2}{R_1^2} C.$$

If C be one standard candle power, we have candle power of source numerically $= \frac{R_2^2}{R_1^2}$.

Another simple photometer suitable for this case is

Rumford's Shadow Photometer. The two sources are set up on the same side of a white screen, before which a vertical rod is also set up. The sources are so arranged as to throw shadows side by side on the screen. Fig. 58 shows a plan of the arrangement. Instead, however, of looking on S_1 as the shadow of source C_1 we rather consider it as a region of the screen illuminated by C_2 only, while S_2 is a region illuminated by C_1 only. The other parts of the screen, with which we are not concerned, are illuminated by both sources.

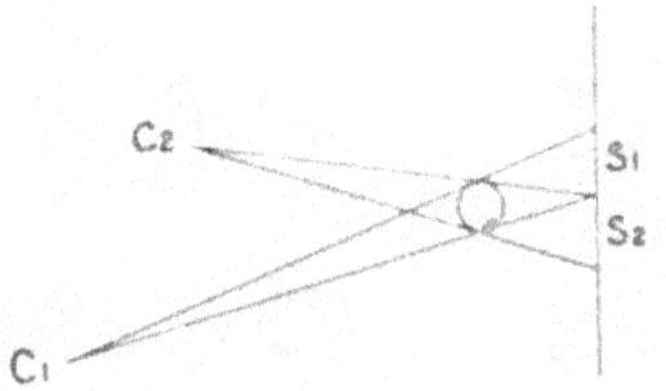

FIG. 58.

When the distances R_1, R_2 from the screen are so adjusted as to give equal illumination over the two regions, we may apply the same calculations as before. This method is not convenient for use with more than two sources on account of the different shadows produced. A little thought will show that the room in which this experiment with the shadow photometer is made need not be absolutely dark, apart from the actual sources used, but in the case of the grease spot this is practically necessary, for any unequal illumination of the two sides by the other light in the room would obviously spoil the result.

A difficulty occurs in balancing illuminations due to different sources, as these sources give out slightly different coloured lights. This will be observed during the experiment with the shadow photometer. The illuminated regions will be perceptibly different in colour. The only way is to adjust till the intensities are judged to be as equal as possible. Strictly, of course, we cannot speak of the absolute equality of the intensities of two illuminations which are different from one another in quality. Try, however, the effect of partly closing the eyes when estimating the equality of the intensities.

The above experiments are all based on the fact that light, and radiation generally, travels in straight lines. Optical instruments depend on the fact that the directions of these lines or rays can be altered by reflection or refraction. Much of the study of optics is therefore concerned with this question of changing the direction of rays.

REFLECTION

The direction of the ray reflected by any surface, plane or curved, is governed by the laws which state that the incident ray, the normal to the surface at the point of incidence, and the reflected ray lie in the same plane, and that the angle of reflection is equal to the angle of incidence, these angles being on opposite sides of the normal.

An ordinary mirror, silvered at the back, is not a simple reflector. Rome rays are reflected from the front surface; some enter the glass, being bent as they do so, and are reflected by the silvered surface. Of these some emerge, others are

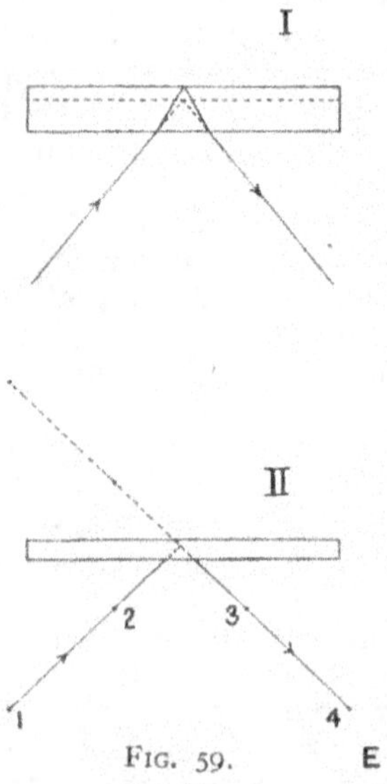

FIG. 59.

internally reflected by the glass surface back to the silvered surface, and so on. The most intense of the beams which emerge is that which has suffered one reflection at the silvered surface—its path is shown in Fig. 59 I. The mirror acts practically for this beam as though it were a simple reflecting surface shown by the dotted line. Strictly, however, the position of this line depends on the angle of incidence of the ray, being a little nearer to the silvered surface for more perpendicular incidence. For accurate optical work such a mirror is useless. A mirror of glass silvered on the front is sometimes used, or simply a polished plate of metal. Or a totally reflecting prism may be used.

Using an ordinary mirror, the equality of angles of incidence and reflection can be demonstrated in the following way: Set up the mirror perpendicular to a sheet of paper pinned on a drawing-board, with its back edge along a line previously drawn on the paper. Set up pins 1 and 2 as shown in the plan II, and, observing from E, set up 3 and 4 in line with the images of 1 and 2. Remove the mirror and pins, and draw the lines 1, 2 and 3, 4. They meet at a point a little in front of the original line. Draw a normal to this through the point of intersection. Measure angles of incidence and reflection with a protractor. Repeat for two or three other angles of incidence. In all these pin experiments the two pins which serve to fix a ray should be at least 3 inches apart. If they are too close together the direction cannot be drawn with any accuracy.

The laws of reflection lead to the conclusion that the image of a point in front of a plane mirror lies on the same normal as the point, as far behind the mirror as the point is in front of it. Set up pin 1 (Fig. 60 I) in front of the mirror, which for this experiment should be a narrow strip, and adjust a tall pin behind the mirror so that its upper portion appears to be a continuation of the image of the first pin viewed from the front of the mirror. Alter the view point from side to side. If pin 2 and image move relatively to one another, readjust 2 until this does not occur however the view point be moved. Then 2 occupies the position of the image of 1. Now, two objects, A, B, Fig. 60 II, viewed from E may seem to coincide, but from E_1 A appears to the left of B and from E_2 to the right. This apparent displacement of an object due to alteration of view point is called Parallax. Suppose B to be the image of 1 as seen in the mirror, and A to be pin 2. Then A must be moved backwards and forwards along the line A E till no parallax occurs as the eye point is varied. Then A and B coincide. The position of the image is thus located. Remove mirror, and join points 1 and 2. Verify the statement given above.

Another method is to trace two or more reflected rays by which the eye sees the image from different points of view, by setting up pins in line with the image, and then produce the lines back till they meet behind the mirror. Thus the image is located (Fig. 60 III).

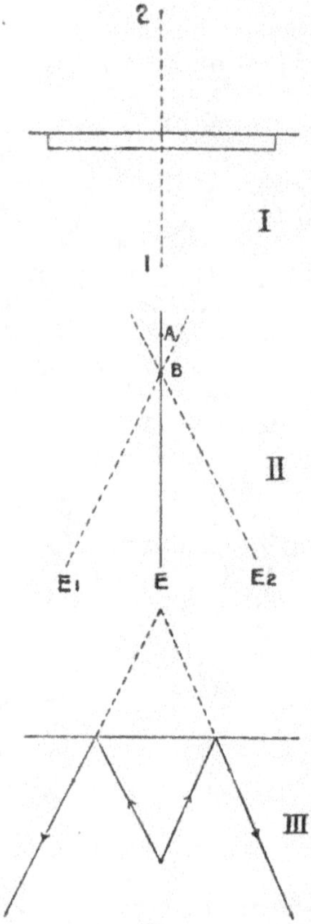

FIG. 60.

THE SEXTANT

A simple instrument, depending on reflection, for measuring the angular distance between distant objects can easily be made. On a triangular piece of wood draw two lines A B, A C (Fig. 61), about 8 inches long, making an angle of 60°. Draw the arc B C and graduate it, or a piece of paper stuck on to the wood, by help of a protractor. But call each degree division 2°, and the whole 60°, 120°. Fit a piece of wood A D so that it can turn about a pivot at the point A. The end D must have a mark to enable its position on the scale B C to be read off. At A fix vertically (on the piece A D) a small mirror, about 1½ inches square, its edge parallel to A D. The point A should be a little in front of the silvered surface, so that incident and reflected rays produced into the glass would meet as nearly as possible at that point. (See Fig. 59 I.) At E set up a piece of wood having an eye-hole at the same height above the base as the centre of the mirror A. At F (A F = A E) set up a mirror similar to A, but with the upper

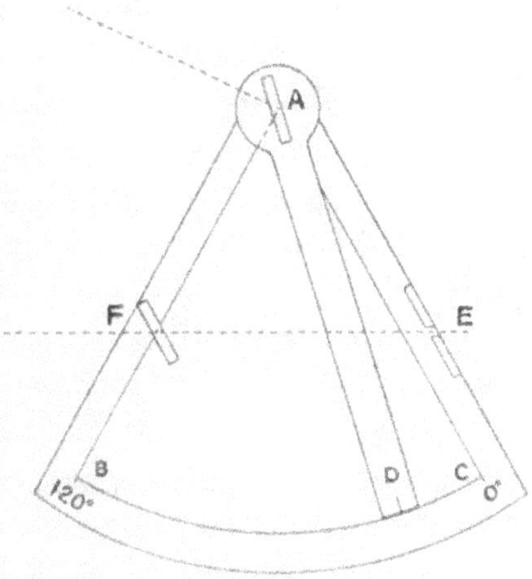

Fig. 61.

half of the silvering scraped away. This mirror must be parallel to the line A C, and must be arranged so that it is at the same height above the base as the mirror A, and it must allow the pointer D to pass under it as far as the 120° division.

A ray passing from A to F exactly over the line A B should be reflected through the hole E if the adjustment is exact. Also when D is at zero, mirror A, being parallel to F, should reflect a ray which falls on its centre, and whose original direction is parallel to F E, along A F. This ray will finally emerge through E. Set the instrument thus, and look through E at a distant point, holding the sextant so as to view the object through F as nearly as possible at the centre of the line dividing the silvered from the clear portion. Observe also the image of the same distant point by reflection at A and silvered part of F, adjusting D slightly if necessary so as to make the image appear to touch the object viewed directly. Note any difference between D and the zero of scale, and in using the instrument always make allowance for this "zero error." If, for instance, D reads 3° when it should read 0°, it will read 23° when it should read 20°, and so on.

Now apply the instrument to determine the angular distance between two objects. Observe one directly through E and F, and turn D till the other is visible by the two reflections as nearly as possible in coincidence with the first. The reading of D, subject to correction for zero error, will give the angle required. If we imagine the rays to be reversed, so as to leave the eye and pass on, one directly through F and the other by two reflections along a line parallel to E F (D being at zero), then as A D is turned through any angle the imaginary reflected ray from A will turn through twice this angle. When it takes up its final position, the angle between it and F E will be twice the angle turned through by A D. Thus the reading of D on the doubled scale will equal the angle between the rays which actually reach the place of observation from the two objects. The slight lateral displacement of the twice reflected ray, from A to F, is immaterial when the objects viewed are distant.

The altitude of the sun at noon may be observed by means of the sextant and an artificial horizon—consisting merely of a dish of mercury—which gives a horizontal reflecting surface. A piece of smoked glass must be fixed over E to protect the eye. The objects viewed are the sun and its image in the mercury. The angle obtained will, of course, be double the altitude of the sun.

Refraction

The direction of a ray refracted at any surface, plane or curved, is governed by laws which state that the incident ray, the normal to the surface at the point of incidence, and the refracted ray lie in one plane, and that the ratio of the sine of the angle of incidence to the sine of the angle of refraction is constant, the rays being on opposite sides of the normal.

The constant ratio referred to is called the Refractive Index, and is denoted by μ, thus $\mu = \frac{\sin i}{\sin r}$, i and r being the angles of incidence and refraction. It is a constant for a given pair of media, separated by the surface at which refraction takes place, and for a given wave length of light.

To determine μ for a specimen of glass in the shape of a block with plane faces perpendicular to its base—which faces need not be parallel—set the block on a sheet of paper on a board and fix two pins to mark out an incident ray, such as 1, 2, Fig. 62 I. Look through at the images of these from the opposite face, and set up two pins, 3, 4, in line with them. Trace with a sharply pointed pencil the two faces of the block, remove and draw in the incident and emergent rays. Join the points where these cut the traces of the block; this line gives the track of the ray inside the glass. For greater accuracy always observe the lower parts of the pins, so as to avoid errors due to the pins being slightly out of the vertical. Consider the first point of incidence. Produce the refracted ray (that in the glass) if necessary, and draw a circle about the point of incidence of, say, 5 cms. radius (Fig. 62 II). Draw the normal to the surface at the point of incidence, and perpendiculars to this from the points where the incident and refracted rays cut the circle. The ratio of the lengths of these perpendiculars is equal to μ, for they are proportional to the sines of the angles of incidence and refraction respectively. Several different angles of incidence should be taken and the constancy of μ verified. If the faces of the block be parallel, the finally emergent ray will be parallel to, but laterally displaced from, the incident ray.

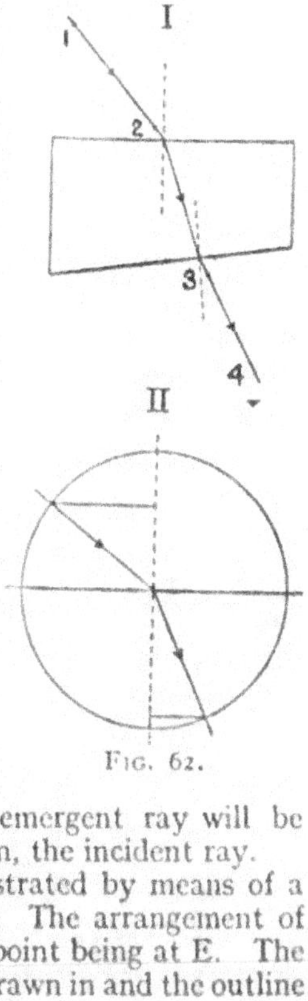

Fig. 62.

Total internal reflection can be illustrated by means of a block of glass, preferably rectangular. The arrangement of pins is illustrated in Fig. 63, the view point being at E. The external parts of the ray can easily be drawn in and the outline of the block traced. Assuming that the laws of reflection hold at the point A, this point can be determined geometrically by the dotted construction, where CD = BC. The straight line joining the first point of incidence with D cuts the reflect-

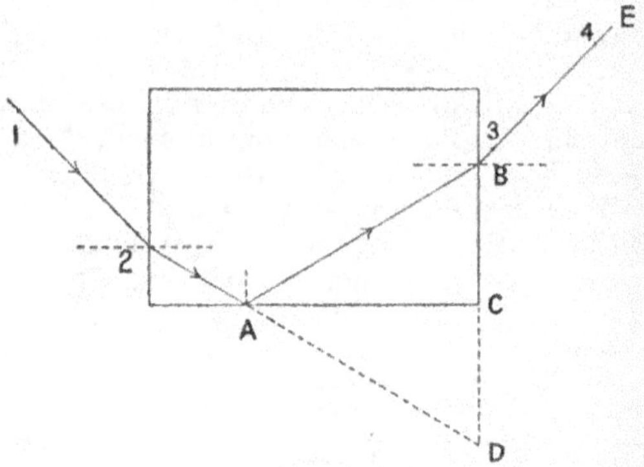

FIG. 63.

ing face at A, and AB gives the track of the reflected ray. The simple proof is left to the reader. It is convenient to use an angle of about 45° with the normal for the incident ray.

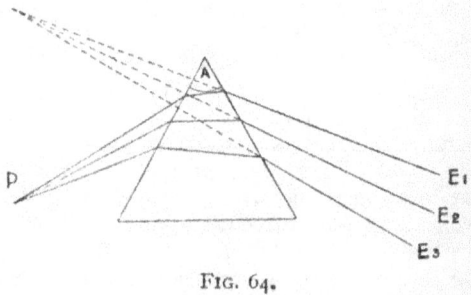

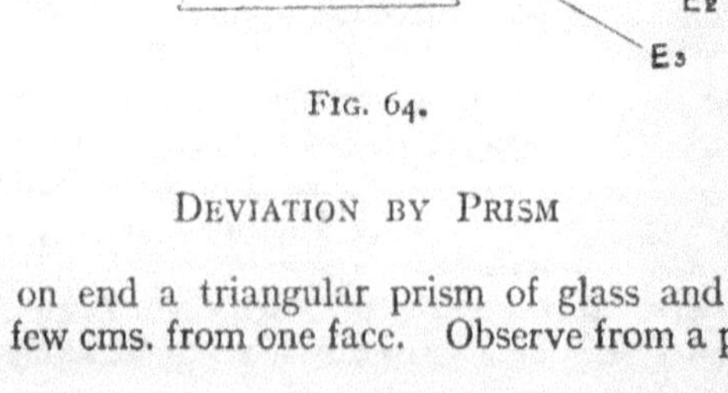

FIG. 64.

DEVIATION BY PRISM

Set up on end a triangular prism of glass and a pin P, Fig. 64, a few cms. from one face. Observe from a point such as E_1 and put in two pins to track the emergent ray. Repeat from one or two other view points E_2, etc. In each case fix an extra pin between P and the prism to track the incident ray. Trace outline of prism, and remove. Draw in the rays, including the portions in the glass, and produce the emergent rays backwards. These will approximately meet at a point which will indicate the position of the image of P as seen through the prism. This, of course, can be located without using the extra pins, but the paths of the rays in the glass could not then be obtained.

The angle between an incident and the corresponding emergent ray is called the Deviation. This has a minimum

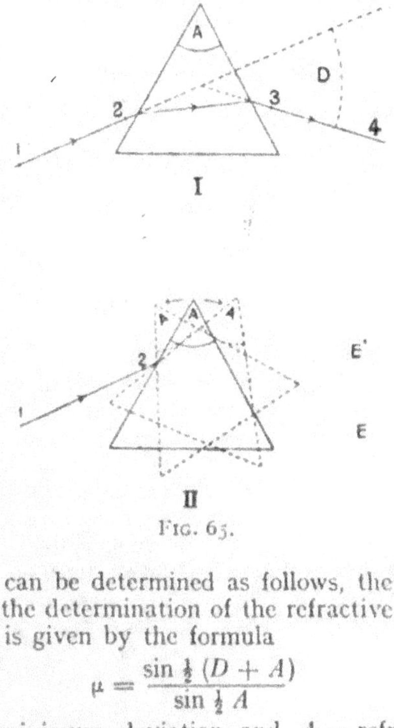

FIG. 65.

value which can be determined as follows, the result being required for the determination of the refractive index of the glass, which is given by the formula

$$\mu = \frac{\sin \frac{1}{2}(D + A)}{\sin \frac{1}{2} A}$$

where D = minimum deviation and A = refracting angle of the prism, marked A in the figure.

Minimum Deviation, 1st Method. Plot a set of incident and emergent rays, using different angles of incidence. Some rays will not emerge from the second face of the prism, but will be totally reflected. But use the greatest range of incident rays possible. Produce the rays, after removing the prism, as indicated in Fig. 65 I, and measure D for each case. Plot a curve using angles of incidence for abscissæ and the corresponding deviations for ordinates. Observe the minimum value of D shown by the curve.

2nd Method. Set up pins 1, 2 (Fig. 65 II) to fix an incident ray and observe these through the prism from a point such as E. Rotate the prism on its base, when it will be found that the emergent ray moves a certain distance in the direction of E′ and then moves back towards E. Follow with the eye, always keeping the images of 1 and 2 in the line of sight, and when the extreme position towards E′ is reached, put in a pin to mark it. Test again by rotating the prism both ways, and then put a fourth pin to fix the direction of the emergent ray. Remove prism, and draw and produce the two rays, and measure the angle between them. This is the angle of minimum deviation. Compare it with that obtained from the graph in Method 1.

Angle of Prism

It is necessary to determine the angle A. Draw two parallel lines (Fig. 66) about 1 cm. apart and place the prism so that its refracting angle lies between them. Set up two pins to indicate the incident ray on one of the lines, and observe the reflections of these in the corresponding face, and fix the reflected ray by two more pins. Do the same for the second face without moving the prism. The angle between the two reflected rays is twice the angle A of the prism. The prism need not necessarily be placed symmetrically with respect to the two lines. The geometrical proof is left to the reader. The accuracy of the determination may be checked by measuring in the same way

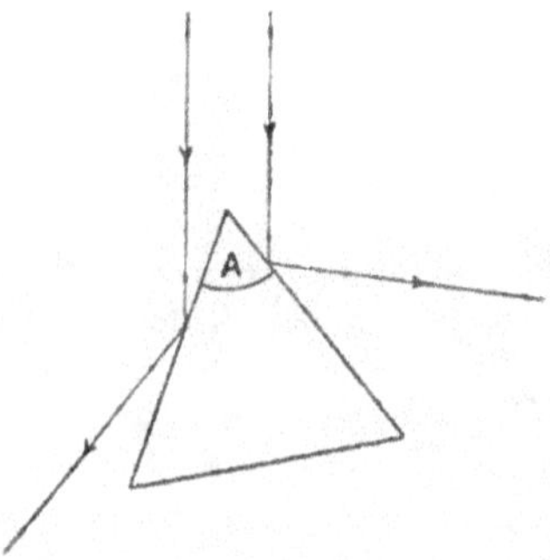

Fig. 66.

all three angles of the prism and adding. The sum, of course, should be 180°.

Having found minimum deviation and angle of the prism, apply the formula to find the refractive index of the glass. Check the result by making a determination of μ by the microscope method as used for a liquid. Make a mark on a flat board and observe it from above by the microscope. Place the prism upright on the mark, and view the image through the upper flat end of the prism. Observe also this end through the microscope, and calculate μ. Before taking readings, see that the microscope is so placed that the extreme points to be observed come within the range of the instrument. It often saves much time to run rapidly over an experiment before commencing accurate observations, in order to see that the various parts of the apparatus used are in proper adjustment. The time spent in careful observation of the mark and its image through the glass in the above experiment is wasted if it be afterwards found that the upper surface cannot be observed without altering the position of the microscope.

Reflection at a Concave Surface

So far we have dealt with reflection and refraction of visual rays only. Set up two large concave mirrors facing one another, several feet apart, in a darkened room. In front of one, opposite its centre, set up a source of light such as an electric lamp, and shade the side remote from the mirror. Adjust till an approximately parallel beam of light is reflected to the other mirror. Find the position of the principal focus F (Fig. 67 I) of the second mirror by holding up a small card to receive the rays. Set up shades, like A. Now light up the room, and replace the lamp by a nearly red hot metal ball, still keeping the shades in position. At F hold the bulb of a thermometer, or some kind of thermoscope. It will indicate the presence of intense radiation. Move it slightly to one side, and little radiation will be indicated. This shows that the dark thermal radiation is reflected in the same way as light.

Again, replace the ball by a watch, taking care that this is placed exactly in the position previously occupied by the lamp. At F hold a funnel connected to a rubber tube which can be brought to the ear. The ticking of the watch will be heard distinctly, but hardly, if at all, when the funnel is slightly displaced. Thus we see that sound is also reflected according to the same laws as light.

Optical Instruments.

These are instruments which cause rays of light, which diverge from certain points, to converge to, or diverge from, other points, or to appear to do so. Take the case of a plane mirror, if we may call it an optical instrument. Rays diverging from A, Fig. 68 I, are made to diverge from A′ or to leave the mirror as though they did so. The eye at E receiving some of these rays sees the "image" of A at A′.

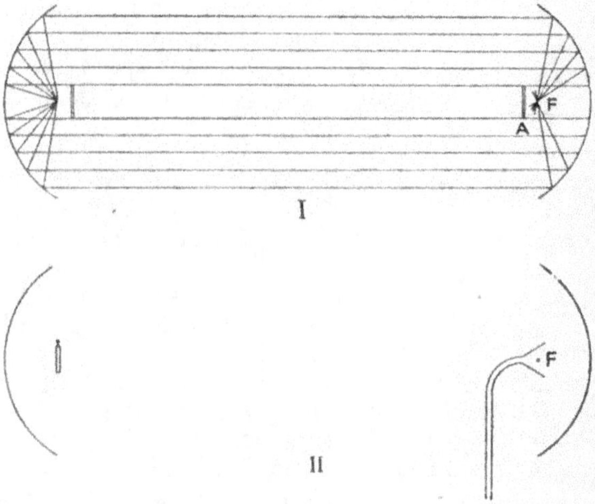

Fig. 67.

Take three points, A B C, representing an extended object. If a real object exactly equal to the image A′ B′ C′ were placed at A′ B′ C′, the mirror might be removed, and if a clear way from E were open, the appearance would be the same to the eye. The function of an optical instrument may thus be said to be the formation of a suitable image of a given object. The eye then looks at this image just as though it were a real object, and the optical instrument were out of the way altogether. Note the relation between A B C and A′ B′ C′.

The lines A A′, B B′, C C′ are parallel to each other. A plane mirror, then, produces an image (which is a virtual one) A′ B′ C′ of A B C, where A A′, B B′, CC′ are parallel. We have already dealt practically with the plane mirror.

Next consider a Magnifying Glass, or Simple Microscope (Fig. 69 I). Rays which actually diverge from A, B and C are made to diverge (or to appear to do so) after they have passed through the lens from A′, B′, and C′. In this case the lines A A′, B B′, C C′ meet at a point (the optical centre of the

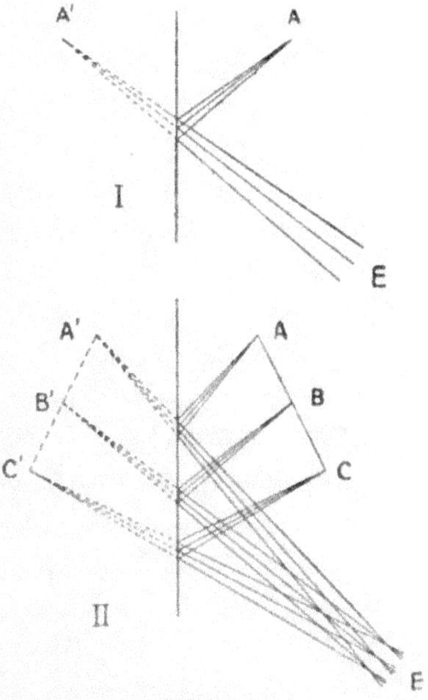

FIG. 68.

lens) instead of being parallel as in the last case. That is also true for a concave lens. A concave mirror (Fig. 69 II) also gives images A′ B′ C′ such that the lines A A′, B B′, C C′ meet at a point (which is the centre of curvature of the mirror). The same is true for a convex mirror. In the cases we have chosen the object and image lie on planes perpendicular to the "optic axes" of the lenses and spherical mirrors.

Suppose an enlarged image is required on the same plane as the object. In this case a single lens or reflecting surface

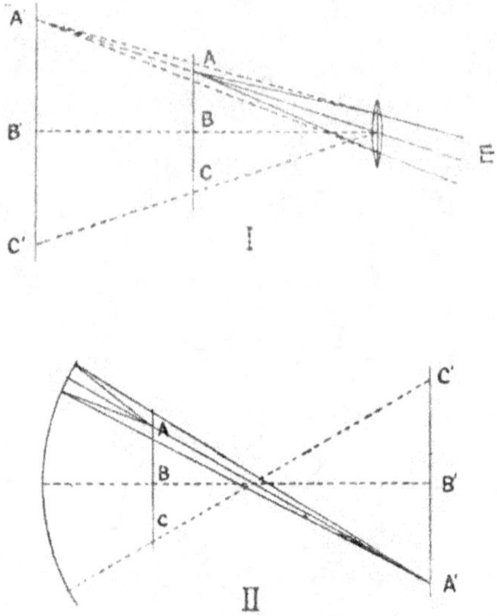

FIG. 69.

will not suffice. Take the case of the compound microscope, and suppose the condition specified to be required Fig. 70.

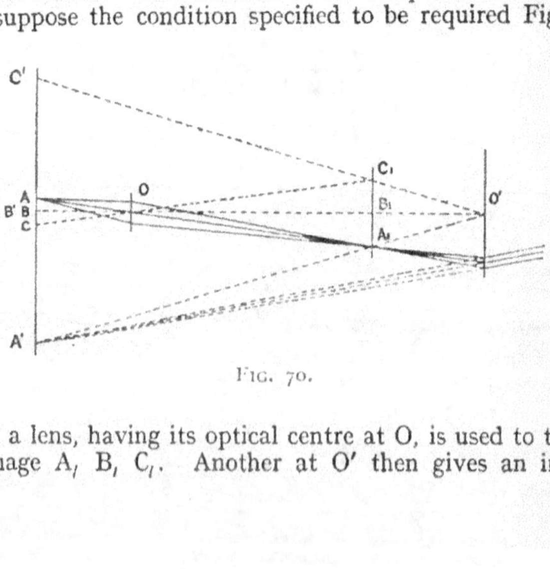

FIG. 70.

First, a lens, having its optical centre at O, is used to throw an image $A_{\prime}$ $B_{\prime}$ $C_{\prime}$. Another at O′ then gives an image

A′ B′ C′ using A, B, C, as its object. In each of the above diagrams a pencil of rays is traced showing the formation of the image A′. In each case we may imagine A′ B′ C′ to be an actual object viewed by the eye with the optical instrument itself removed. The rays which actually enter the eye after passing the instrument are just the same as if they came from the supposed object at A′ B′ C′, no instrument being present.

In the compound microscope, even if the image were not required to be in the plane of the object, but in such a position as shown in Fig. 71, two lenses would be required practically, the angle being too wide for a real lens to produce, or to enable the rays to enter the eye even if produced.

The points A, A′, etc., are called Conjugate Foci in the case of a lens or spherical mirror. The law connecting their distances from lens or mirror is given by

$$\frac{1}{v} \pm \frac{1}{u} = \frac{1}{f},$$

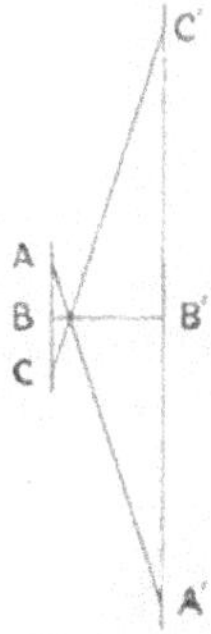

FIG. 71.

the + sign relating to the mirror and the — to the lens. In the formula u denotes distance of object from lens or mirror, v distance of image, and f is a constant, depending on the particular lens or mirror, called the focal length. All these lengths are measured from lens or mirror in the direction from which the incident light comes. It is evidently of importance to know the value of f for each lens and mirror used. In the case of a mirror, f is equal to half the radius of curvature, r, so determination of r may be made if more convenient. In the case of a lens

$$\frac{1}{f} = (\mu - 1)\left(\frac{1}{r_1} - \frac{1}{r_2}\right),$$

where r_1 is the radius of curvature of the surface of the lens on which the light is incident, and r_2 of that from which it emerges, both r_1 and r_2 being measured from the lens in the direction specified above. Thus if these radii were known, and also the refractive index of the glass, f could be calculated.

FOCAL LENGTHS OF LENSES

I. *Convex Lens.* Use an optical bench, or long graduated bar fitted with sliding stands for lenses, etc. (Fig. 72). For object a suitable arrangement consists of a piece of fine wire gauze fixed in a hole flush with the front of a small lamp box. The front itself should be whitened so as to act as a screen in the same plane as the object. A measuring rod of known

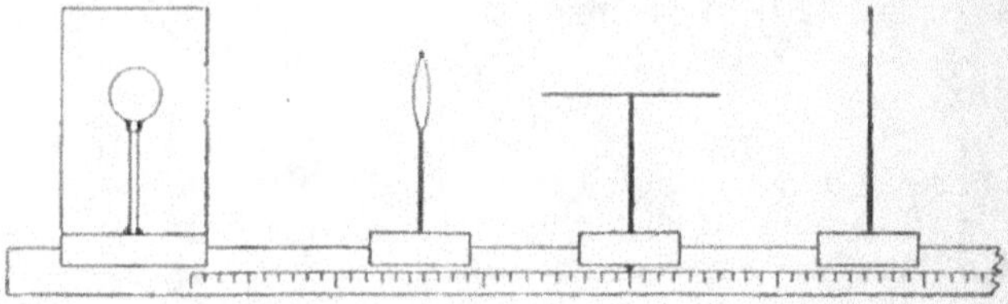

FIG. 72.

length attached to a sliding stand which has a pointer moving over the scale on the graduated bar is also necessary.

Arrange the illuminated object, lens, and a separate screen, so as to obtain a clear image of the gauze. The distance from object to screen must be at least four times the focal length (numerically—the focal length of the convex lense is negative)

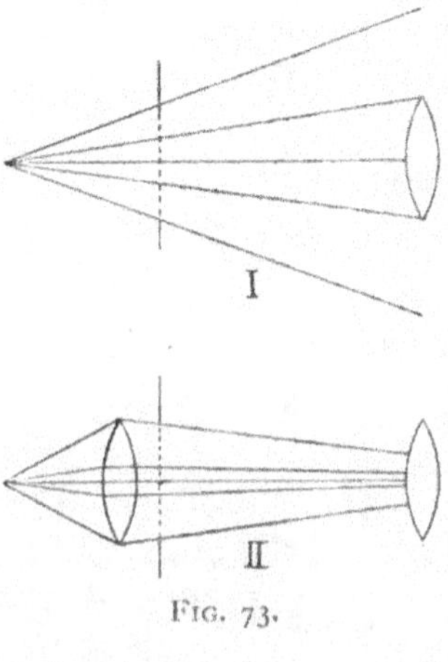

FIG. 73.

or no clear image can be formed. If they be this distance apart the lens will be midway between them. If farther apart, two possible positions of the lens can be found, one giving an enlarged and the other a reduced image. If the former, the arrangement is that of the optical lantern. If possible, try the effect of interposing a lens (called a condenser when thus used) between lamp and gauze, inside the box, of such focal length as to concentrate the beam of light on to the other lens. Fig. 73 I and II show the arrangement without and with the condenser.

To determine *f*, measure distance from lens to object and from lens to image. Use the measuring rod, bringing it first into contact with the gauze and then with the surface of the lens next the gauze, and take readings of its pointer. The distance required is equal to the distance moved by the rod plus its own length. Repeat on the other side of the lens. Half the thickness of the lens may, for greater accuracy, be added to each measurement. This can be deduced from the two readings of the pointer taken when the rod touched the faces of the lens. A little thought will show how to do this ; perhaps the help of a little diagram will be useful. (Always make plenty of simple diagrams to illustrate the various points which arise in the study of lenses, etc.) Suppose distance of object from lens to be 20 cms., and distance of image on the other side of lens 30 cms. Substitute in the formula $\frac{1}{v} - \frac{1}{u} = \frac{1}{f}$ * the values 20 for u and —30 for v, for the image is on the negative side of the lens. Then $\frac{1}{(-30)} - \frac{1}{20} = \frac{1}{f}$, from which f is found to be —12, i.e., $f = -12$ cms. Repeat for the second position of the lens, object and screen being as before, and also for several different distances of object to screen.

Another method, which occupies a shorter length of the bench, is as follows. Set up object box, lens, and a plane

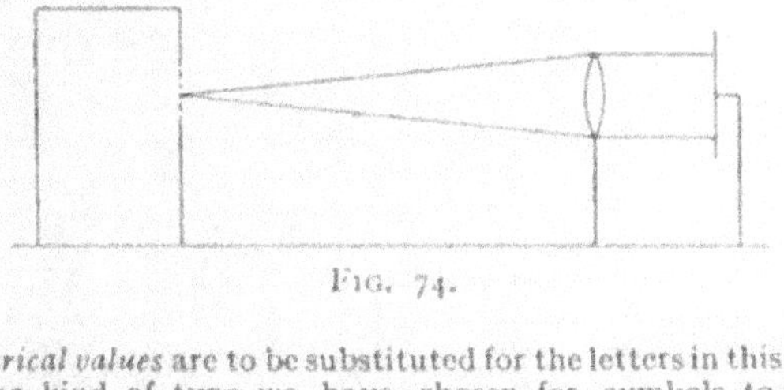

FIG. 74.

* *Numerical values* are to be substituted for the letters in this equation, and so the kind of type we have chosen for symbols to represent numerical values of physical quantities is here used. This need not disturb the student.

mirror in line, as in Fig. 74. Adjust the lens till an image is thrown on the box close beside the object. In this case the rays which return from the plane mirror practically retrace their original paths, except for a necessary displacement to enable the image to fall a little to one side of the object. This can only be the case if the rays fall on the mirror in a parallel beam. Now consider only the returning rays. They, being parallel, may be considered as coming on to the lens from an infinitely distant object and forming an image on the box. The distance from lens to box is then equal to the focal length, for since u is infinite we have

$$\frac{1}{v} - 0 = \frac{1}{f} \quad \text{or} \quad v = f.$$

The focal lengths of a long and a short focus lens should be determined; they will be required for subsequent use in the setting up of optical instruments.

II. *Concave Lens.* Combine with a short focus convex lens of known focal length, and use the pair as a convex lens. The focal length f of the combination is given by

$$\frac{1}{f} = \frac{1}{f_1} + \frac{1}{f_2},$$

where f_1, f_2 are the focal lengths of the separate lenses; hence that of the concave can be found. If a suitable convex be not available, the following method may be used. Set up object, a convex lens, and screen, on which obtain an image. Interpose the concave lens, backed by a plane mirror (Fig. 75),

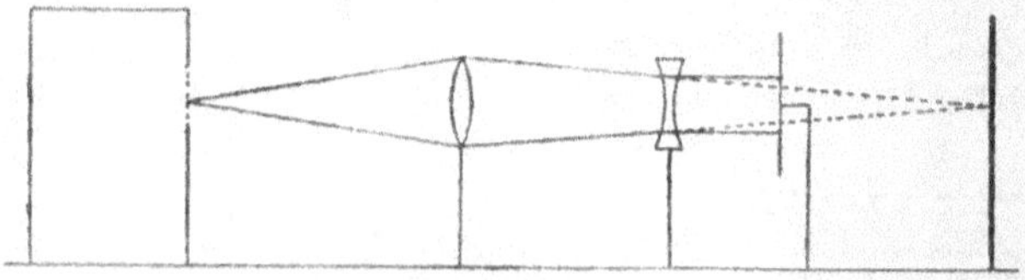

FIG. 75.

and adjust it so that an image is thrown back to the object. Then the rays, after leaving the concave lens on their outward journey, must be parallel. (That is the set of rays which originally diverge from one point of the object.) The distance from the concave lens to the screen on which the convex lens threw the original image is equal to the focal length of the concave lens. For a system of rays converging to a point falls on the concave lens and is rendered parallel. That point is therefore the principal focus of the lens (Fig. 76).

The focal length of a short focus concave lens will be required for subsequent work.

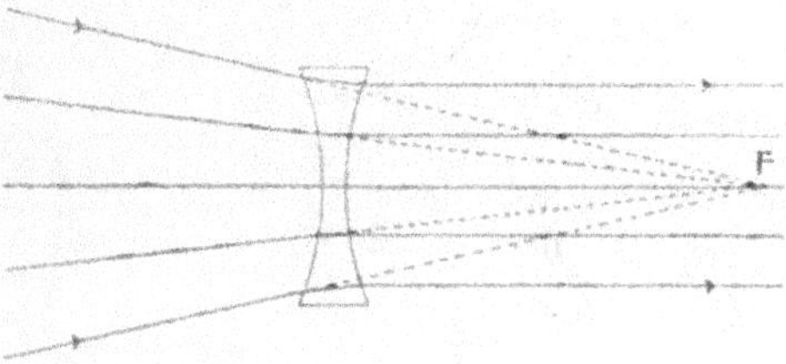

FIG. 76.

III. *Concave Mirror.* Set up the mirror facing the object and at such a distance as to form an image close to the object. Then the distance of either from the mirror is equal to the radius of curvature, for the rays practically retrace their own paths and must therefore fall normally on the mirror at each point. The slight displacement necessary to cause the image to fall to one side of the object hardly affects the distance. Object and image at different distances may be used by

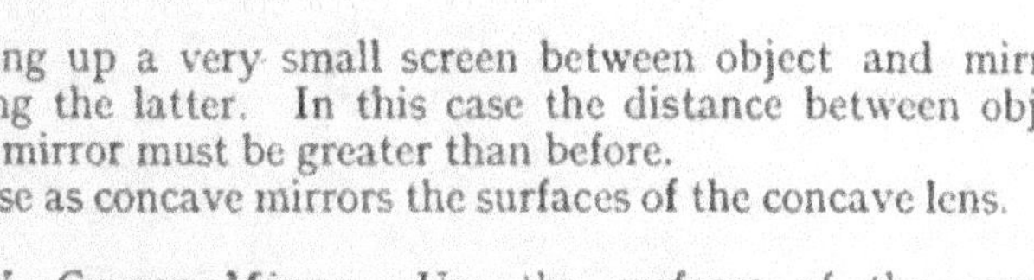

FIG. 77.

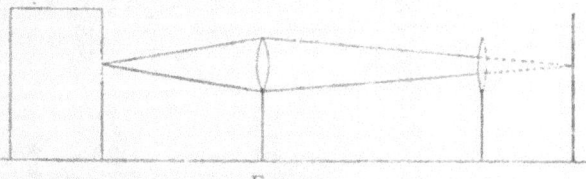

setting up a very small screen between object and mirror, facing the latter. In this case the distance between object and mirror must be greater than before.

Use as concave mirrors the surfaces of the concave lens.

IV. *Convex Mirror.* Use the surfaces of the convex lenses, or at least of one of them. Set up an auxiliary lens to throw an image on to the screen. Interpose the mirror (Fig. 77) at such a distance as to throw back an image to the plane of the object. Then the rays reflected must retrace their own paths, and therefore they were incident normally on the convex surface. Therefore these incident rays produced must meet at the centre of curvature of the surface. But they met at the screen before the surface was interposed. Therefore the distance from the surface to the screen is numerically equal to the radius of curvature. The radius is negative in this case, being on the side opposite to that from which the incident light arrives.

Check the values of radius of curvature in all cases by means of the spherometer, as explained below, and calculate the refractive index of the glass of which each lens is made by means of the formula $\frac{1}{f} = (\mu - 1)\left(\frac{1}{r_1} - \frac{1}{r_2}\right)$, taking care to apply the correct signs to the numerical quantities substituted.

The Spherometer

This is a little three-legged instrument (Fig. 78) with a central screwed pin having a graduated head—frequently

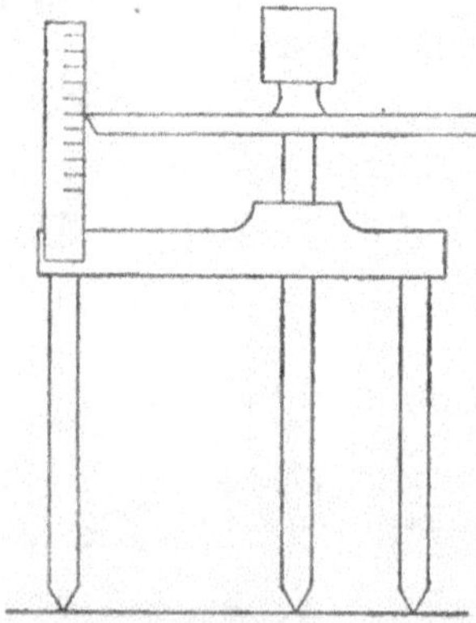

Fig. 78.

divided into 100 parts. A side mm. scale enables the value of the pitch of the screw to be determined. For this purpose the screw should be rotated a definite number of times, and the traverse noted on the side scale. The pitch will probably be 1 mm. or $\frac{1}{2}$ mm.—it should have an exact value in mms. or else it would be impossible to estimate it correctly by this means.

To use the instrument place it on a piece of plate glass, and screw the pin till it just touches the plate. If screwed slightly too far, the spherometer can be rocked by the fingers; with a little practice great accuracy can be attained in the adjustment. Read the graduated head, and place the spherometer on the spherical surface. Screw up or down as

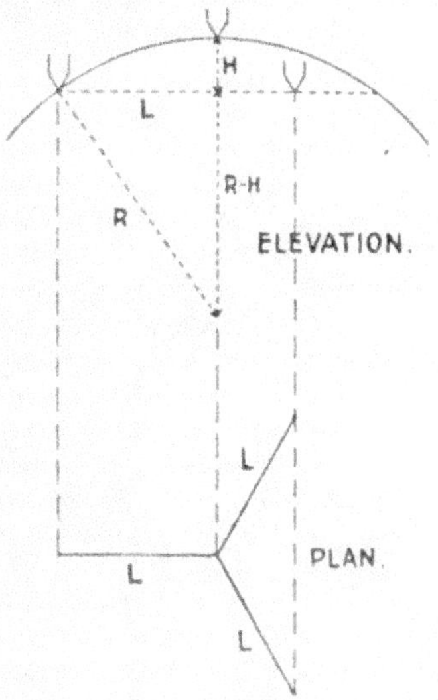

Fig. 79.

the case may be, till all four feet accurately touch the surface. Count the number of turns, and determine the odd hundredths of a turn by again reading the head. Thus the number of turns is obtained to two decimal places. Convert this to mms. of traverse by a knowledge of the value of one turn of the screw as already found. The distance H, Fig. 79, is thus determined. Now R being the radius of curvature required,

$$R^2 = (R - H)^2 + L^2$$

or

$$R = \frac{L^2 + H^2}{2H}.$$

To determine L, level the instrument and press gently the four points on to a sheet of paper. Remove and measure the distance from centre mark to each of the others. Take the mean value if these are not identical, in this way practically eliminating any error due to slight want of symmetry of the instrument.

Work out on similar lines the expression for R for a concave surface.

The spherometer can also be used for measuring the thickness of a small object, which can be placed on the plate of glass under the central screw, and for other purposes for which fine measurements of a similar kind are required.

Optical Instruments.

I. *Astronomical Telescope.* For this experiment a smooth metal rod or tube set up on a stand by means of a lug and pivot may be used. Frames in which the lenses can be fitted, provided with springs as shown in Fig. 80, to clip the rod

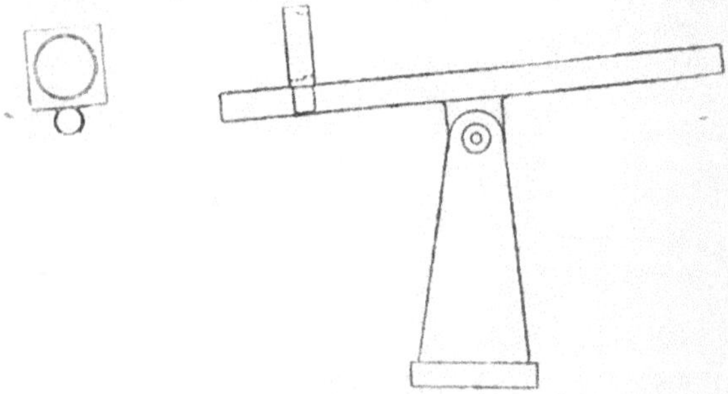

Fig. 80.

complete the apparatus. Set up the short focus convex lens at one end to form the eye-piece, and the long focus one at a distance numerically equal to the sum of the focal lengths for the object-glass. Observe a distant object, and make any slight adjustment which may be necessary. Draw to scale a diagram showing the path of a pencil of rays (as from one point of the object) to the eye, and produce the rays, which enter the eye, backwards to indicate the position of the image observed. Since the object is distant, the original pencil from one point of it is practically parallel, and, if the adjustment be as above described, the image is also distant, so the rays entering the eye are parallel to one another. The diagram is shown in Fig. 81. The dotted line A B is a con-

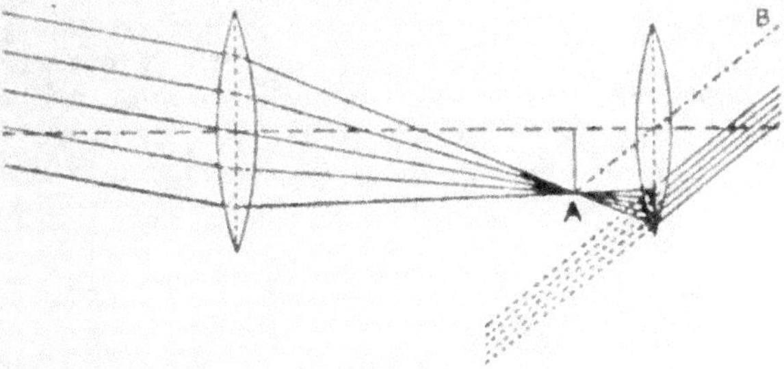

FIG. 81.

struction line necessary to give the direction of the emergent pencil. If one of the actual rays from A passed through the centre of the eye-piece it would be undeviated, and A being in the " focal plane " of the eye-piece all other rays from that point would emerge parallel to this one. Hence the reason for the construction adopted.

II. *The same used as an Observing Telescope* for near object. Draw out the telescope by moving the object-glass farther along the rod, till the object is clearly visible. For the sake of variety arrange the lenses so as to produce the image at a given distance, say 50 cms., from the eye-piece. This can be done approximately by setting up a pointer of some kind at the required distance, and looking at this with one eye while observing the image through the telescope with the other, and focusing the image by moving the object-glass along the rod. When both appear in focus at once, their distances from the eyes will be, at least roughly, the same. This method, of course, cannot be applied where accuracy is necessary, but there is no special point in attaining great accuracy in the distance of this image. Draw to scale a diagram, using the known distances of object, and final image, and focal lengths of the lenses. But in the scale drawings lateral dimensions should be increased, or they will be too small compared with the axial dimensions. (See Fig. 82.) Draw object, object-glass, and image I′ thrown by it. To find position of eye-piece take a subsidiary diagram. Distance of final image is known. Its size is immaterial for the present diagram. Draw image (inverted) and eye-piece, 50 cms. apart, to scale. Where is the object which would throw this vertical image? It must be on the line joining image to centre of lens, or on this line produced. (We are now considering one point of the image and the corresponding point of the object.) Again, one of the rays which seems to come from the image passes through F. This must really have entered the eye-piece parallel to the axis, from the object. The object is therefore as shown in the figure. But this is really I′. Thus the distance of the eye-piece from I′ is

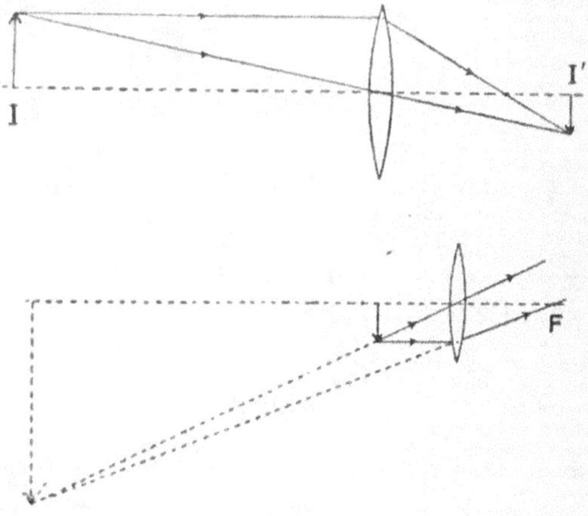

FIG. 82.

determined. The same result could be obtained by calculation, for if u be the distance of I′ from eye-piece and v the distance of its image, i.e., $v = +50$ cms., we can, since f is known, determine u by means of the equation $\frac{1}{50} - \frac{1}{u} = \frac{1}{f}$.

Compare the distance from object-glass to eye-piece in the diagram with that found in the actual experiment. Finally draw a pencil of rays as in the case of the astronomical telescope.

III. *Galilean Telescope.* For object-glass use the same

lens as before ; for eye-piece the concave lens. The telescope will be short, and will give an erect image. Use it for an object at a moderate distance. Repeat the diagrams, making the necessary modifications for this case. These should be worked out by the reader himself. Compare the length of the actual telescope with that obtained in the diagram.

IV. *Compound Microscope.* Set up a piece of gauze in one of the lens holders on the stand. Use the short focus convex lens for object-glass and the long focus convex for eye-piece. Note how the magnification varies for different distances between the two lenses. For each distance chosen, focus by moving the object backwards and forwards along the rod. This is equivalent to keeping the object fixed and moving the microscope, as is usual in actual practice.

Determine the magnifying powers of the observing telescope and of the compound microscope, or of actual instruments of these types. For the former set up a large and coarse scale; observe this directly with one eye and through the telescope with the other, focusing till both are clear. The two appearances must be superposed. Count the number of the small divisions (seen directly) for a given number of the magnified divisions. The ratio gives the magnifying power for objects at that distance from the telescope.

For the microscope it is necessary to arrange for the final image to be at the least distance of distinct vision from the eye. Set up a scale A, Fig. 83, slightly to one side of the axis

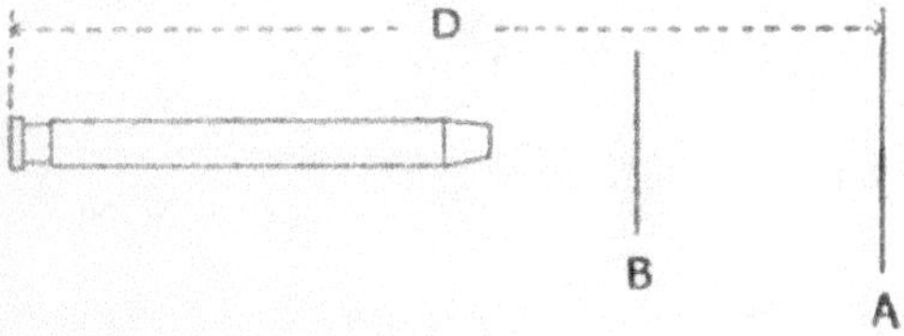

FIG. 83.

of the microscope, at this distance from the eye-piece, to be viewed directly. Set up another scale B opposite the microscope and move it backwards and forwards till its image is clearly in focus at the same time as the directly viewed scale. Move the latter sideways, if necessary, to cause the appearances to be superposed. Count, and take the ratio as before. Repeat with the microscope drawn out to different lengths. In all cases the final image must be at the least distance of distinct vision. It must not be near enough to cause the slightest fatigue to the eye, as this would soon more than counterbalance any advantage derived from the resulting increase of magnification.

The Spectrometer.

This comprises a collimator (A, Fig. 84), telescope B, and a

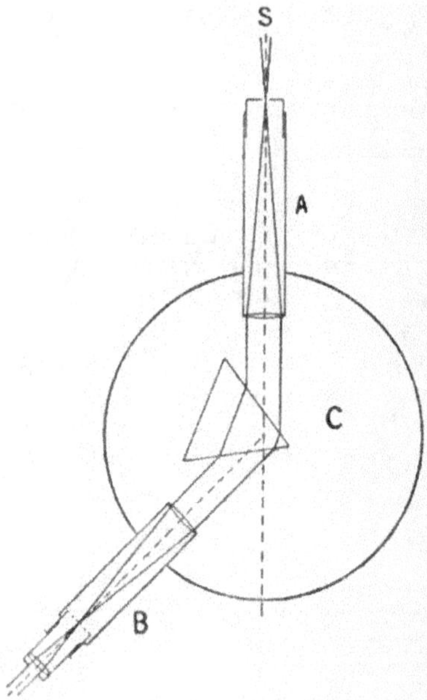

Fig. 84.

table C for holding a prism or other piece of apparatus. The telescope and table are capable of independent rotation about a central vertical axis, and are provided with scales by which the rotations can be measured. Light from a source S passes through a narrow slit at one end of the collimator, and is rendered parallel by a lens at the other end. This parallel beam may be split up into a series of parallel beams by a prism, but if the original light was monochromatic only one beam will emerge after refraction by the prism. In any case, when a parallel beam enters the telescope it is brought to a focus on the plane of the eye-piece cross-wires. An image, in fact, of the slit is there produced. This is enlarged and thrown back by the eye-piece, together with the image of the cross-wires.

To adjust the apparatus, remove the prism, and focus the eye-piece so that the cross-wires can be seen distinctly. Then focus the telescope on a distant object, that is, one from each point of which rays, practically in a parallel beam, reach the telescope. The image of this thrown by the object-glass must lie in the same plane as the cross-wires. Ensure that this is so by using the parallax method—move the eye as far as possible from side to side and focus till no relative motion of image and cross-wires is observed. Now bring the telescope into line with the collimator, and adjust the latter by moving the slit in and out till a sharp image of it is visible on the cross-wires. Use again the parallax method. The collimator slit is now in the principal focus of the lens, and the light crossing the table is parallel—otherwise it would not be focused on the cross-wires of the telescope, the latter having been adjusted for parallel light only. Other adjustments may be necessary with particular instruments, but need not be described here.

Set up a sodium flame—easily made by clipping with a piece of tin a square of asbestos soaked in brine to a Bunsen burner, in the position shown in Fig. 85, several inches from the collimator slit. Place the flame edgeways to the collimator, in order to obtain the most intense illumination. Set up the prism so as to reflect part of the beam on either side (Fig. 86 I). Observe the images of the slit, produced by reflection, with the telescope, adjusting in each case till the image is exactly central with respect to the cross-wires. Half the angular distance traversed by the telescope between its two positions gives the angle of the prism.

Next turn the prism till an image by refraction can be observed (Fig. 86 II). Find this first by eye, and then bring the telescope into position. Slowly rotate the table and prism, first one way and then the other, and observe that the deviation decreases to a minimum and then increases again. Set the telescope to this minimum position. When correctly set, the image of the slit, as the table is rotated, will just move up to the centre of the cross-wires and then move back again. Take the reading of the telescope, and then arrange the prism so as to deviate the beam towards the right, and take a new reading of minimum deviation. The angle of minimum deviation is half the angle turned through

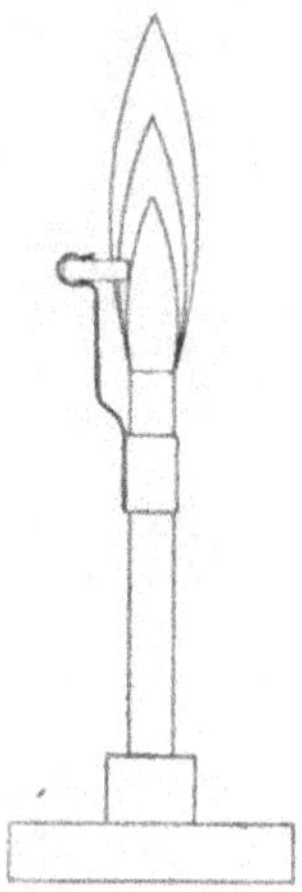

FIG. 85

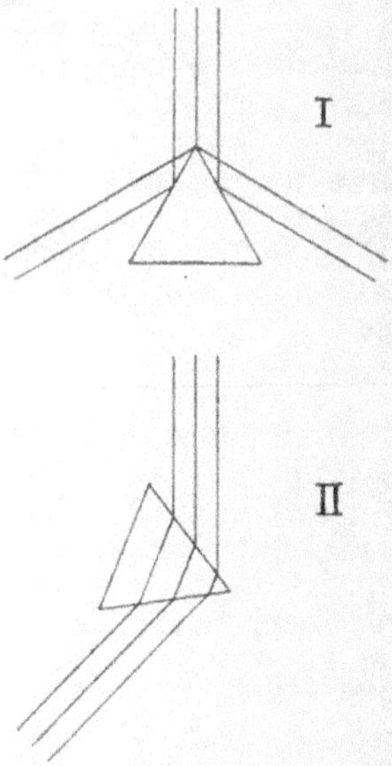

FIG. 86.

by the telescope between its two positions. Calculate μ for the material of which the prism is made.

A hollow prism may be used for the determination of μ of liquids. The plates of glass (if of uniform thickness) forming the faces of the prism do not affect the directions of the rays; the deviation is caused only by the liquid itself. A number of prisms, solid and liquid, of the same angle but of different materials, should be used to demonstrate the different values of the deviations produced. At least, flint glass, crown glass, and water should be tried.

Dispersion

Substitute white light, or ordinary gas light, for the sodium light. Instead of a single image of the slit, a continuous spectrum will be observed through the telescope. It may be looked upon as an infinite series of images of the slit in various colours ranged along a line. Touch the flame with a piece of common salt or a wire dipped in brine—the yellow sodium line will appear strongly marked at its proper place in the spectrum. Test roughly the length of the spectrum for the different prisms.

Using the flint glass prism, observe the line spectra for several substances, colouring the Bunsen flame in the following way: Make a very loose bundle of several thin iron wires, bend it, and fit it into a small test tube, as shown in Fig. 87.

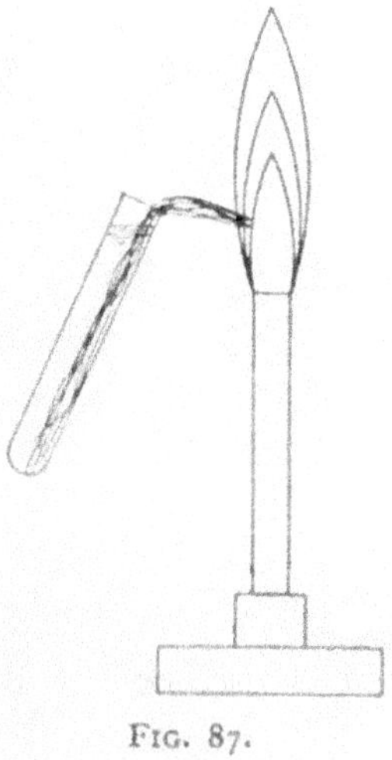

Fig. 87.

Fill the tube with a strong solution of, say, strontium chloride in water. Wet the overhanging part of the bundle, which should not be too long, say about an inch, and bring it into the flame. A gradual flow will be kept up. Slightly lean the tube if necessary. Use a separate arrangement for each salt. Determine minimum deviation for several lines belonging to different substances, including sodium, and tabulate the values, noting down the colour of each line. Look up in the tables the wave length of each of the lines, and plot a graph showing wave lengths and minimum deviations.

Another method is to fix the prism in the position of minimum deviation for sodium, and, without moving this, take readings for each of the lines observed. Then plot wave lengths against these readings. Now take an unknown solution which gives a line spectrum. Observe the minimum deviation (or the reading, if the second method be used) for each line, find from the curve the corresponding wave lengths, and then search the tables for a substance whose lines correspond to those wave lengths. This is a simple case of spectrum analysis.

MAGNETISM

Verification of the Law of Force due to a Magnet; and Determination of the Moment of a Magnet, and of the Horizontal Intensity of the Earth's Field

SUPPOSE a bar magnet to be placed on a horizontal table, its negative or south pole pointing due north. At any point distant r from the centre of the magnet along the axis, in either direction, the force due to the magnet is directed towards the south. This force, in dynes acting on unit pole, is given by

$$F = \frac{2Mr}{(r^2 - l^2)^2},$$

M being the moment of the magnet, and l half the distance between its poles. The horizontal force due to the earth, acting as a magnet, in dynes per unit pole, is denoted by H, and is directed towards the north. Assuming then that F is greater than H for points near the magnet, it follows that at two points along the axis F and H will be equal and opposite. At those points, called neutral points, the resultant force is zero. They are symmetrically situated with respect to the magnet. If M and H be known, the positions of the neutral points can be calculated. It is, however, important for other purposes to be able to determine M and H. But before making this determination it will be well to verify the law of force given above. The experiment will also help towards the determination of M and H.

Use a magnetometer—a small pivoted magnetic needle enclosed in a box placed at the centre of a long graduated rod, the needle being provided with a double-ended pointer which moves over a circular scale. Place the rod due magnetic east and west by adjusting till the needle is perpendicular to the rod. Set the magnet on the rod with, say, its positive pole towards the needle. Let the angle of deflection of the latter be θ. To determine this, both ends of the pointer must be read so that errors due to imperfect centring of the pivot may be avoided. The force acting on a pole, say m, of the needle, due to the magnet, is $F\,m$ (in dynes), and the force perpendicular to this, due to the earth's magnetism, is $H\,m$. Since the needle takes a position in the direction of the resultant of these forces it can easily be shown that $\tan\theta = \frac{F}{H}$. But practically it is necessary to take 4 positions of the magnet—magnet to E. of needle, its + pole to E.; ditto with + pole to W.; magnet to W. of needle, its + pole to E.; ditto with + pole to W.: the centre of magnet being in all cases at the same distance r from centre of needle. In each case both ends of the pointer must be read. Thus 8 readings of θ are made, of which the mean is to be taken. The reversal of the magnet, end for end, reverses the deflection of the needle; the mean deflection is thus independent of small errors in the zero reading of the needle. Changing the magnet from one side of the needle to the other eliminates any error due to the magnetic centre of the magnet not coinciding with the centre of the bar itself. Several distances $r_1\,r_2\,r_3$ must be taken, and the final value of θ for each determined, say $\theta_1\,\theta_2\,\theta_3$. The length (magnetic) of the magnet, or distance between the poles, may be obtained by placing a small compass needle close to the ends of the magnet in several positions. In each position the needle will point approximately to the pole. Half the distance between poles is the l of the formula.

Now $$\frac{F}{H} = \tan\theta \text{ or } \frac{2Mr}{(r^2 - l^2)^2 H} = \tan\theta,$$
thus numerically
$$\frac{(r^2 - l^2)^2 \tan\theta}{r} = \frac{2M}{H} = \text{constant}.$$

To verify the law, fill in the numerical values for r_1, θ_1, for r_2, θ_2 and for r_3, θ_3, and test the constancy of the results. Take the mean, if the results are not quite identical, and divide by 2. The result is the numerical value of $\frac{M}{H}$.

Another experiment is necessary; this will give the value of $M\,H$. It is known as the vibration experiment, the previous one being called the deflection experiment. Use the same magnet at the same station as before, so that the M and H in this experiment may be the same as those in the last. Suspend the magnet by means of a paper stirrup and unspun silk fibre so that it can oscillate in a horizontal plane. If the magnet be of rectangular section, two of its sides must be in horizontal and the other two in vertical planes (Fig. 88). If the fibre be not at first free from torsion, this must be removed by putting in the stirrup a bar of non-magnetic material equal in weight to the magnet. The stand to which the fibre is attached should then be turned till this bar, when at rest, points N. and S. Then, when the magnet is substituted, it will point N. and S. while the fibre will remain free from torsion. Set the magnet swinging over a small angle. This can be done, if the magnet be covered with a glass shade, as it should be, by bringing up another magnet, which, however, must be removed to a great distance before the experiment is continued. Count and time a number of complete oscillations, say 20, always timing at the instant the magnet crosses its zero position, which should be suitably marked on the base or on the glass shade. The time of swing for one complete oscillation is given by

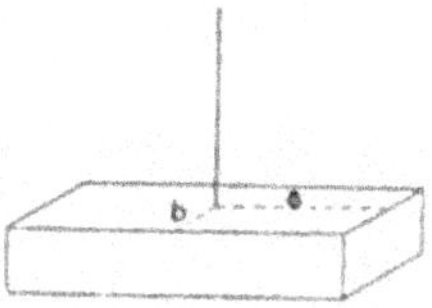

FIG. 88.

$$t = 2\pi\sqrt{\frac{I}{MH}}$$

where I is the moment of inertia of the magnet. If this be rectangular and supported as shown in the figure, $I = \text{mass} \times \frac{a^2 + b^2}{3}$. If cylindrical, of length l and radius r, $I = \text{mass} \times \left(\frac{l^2}{12} + \frac{r^2}{4}\right)$. The value of $M\ H$ can easily be obtained, and by combining this with $\frac{M}{H}$ from the last experiment M and H can be calculated.

FORCE AT POINTS ALONG THE AXIS

We are now in a position to determine the actual force at any point along the axis of the magnet, for, M being known, F can be calculated from the formula $F = \frac{2Mr}{(r^2 - l^2)^2}$ for any distance r from the centre. Calculate for several distances and plot as in Fig. 89. The curve is independent of the position of the magnet with respect to the earth, but we suppose the magnet to be along the magnetic meridian with its S. pole towards the north. Now, H also being known, its value can be plotted—below the axis, for its direction is contrary to that of F. The neutral points are shown by crosses at the points where F and H are equal in magnitude. They can most easily be found by drawing a horizontal line as far above the r-axis as H H is below it. The points where

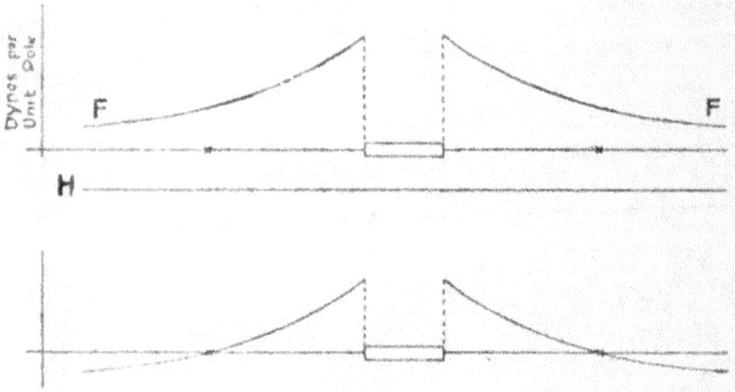

FIG. 89.

this cuts F F correspond to the neutral points. The resultant of F and H can be plotted by reproducing F F and subtracting from each of its ordinates the value H. This amounts to drawing F F to an axis raised H above the original axis, or simply to lowering F F by an amount H. The result is shown in the lower figure. The force at places between the magnet and the neutral points is directed towards the S., beyond these points towards the N.; and the force is zero at the points themselves.

Other neutral points exist for different positions of the magnet. Variations of the above exercises can easily be devised.

Lines of Force

Place the magnet on a sheet of paper pinned to a board, in the position used above. Plot lines of force around it by means of a small suspended needle. Mark the paper immediately under both ends a and b of the needle, and move on the latter till a coincides with the mark previously corresponding to b. Mark the new position of b, and so on. In this way map out the whole horizontal field around the magnet. Note specially the configuration of the lines near the neutral points. The approximate positions of these will be evident on the diagram. Compare their positions with those found by the last method.

Stretch a sheet of paper on a frame and place this horizontally over, and just touching, the magnet. Sprinkle iron filings all over this paper, not too thickly, and tap gently. The filings will indicate the general directions of the lines of force, at least near the magnet where the force is sufficiently strong. Fields should be plotted for other positions of the magnet.

Resistance

DETERMINE the resistances of wires of several materials, such as copper, iron, manganin, and nickel silver, by the Wheatstone bridge method. Many adaptations of the method are in use, based on the following principles. Four resistances *P*, *Q*, *R*, *S*, one unknown (say *R*), are arranged in a loop. The junctions are joined to cell and galvanometer as shown in Fig. 90, keys being interposed

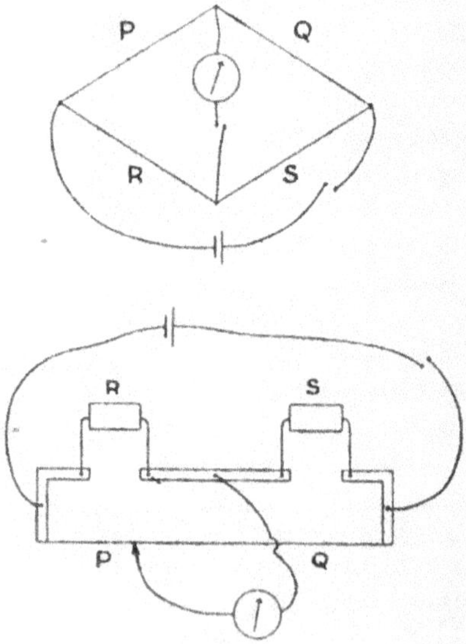

FIG. 90.

in these branches. Cell and galvanometer may be interchanged without altering the final result. Using the arrangement shown, it is only necessary to note that the cell is connected to the junctions of *P* and *R*, and of *Q* and *S*, and the galvanometer to those of P and Q, and of R and S. Once this simple arrangement is understood no difficulty should be found in comparing any ordinary form of Wheatstone bridge with the diagram and connecting it up for use. In some designs S may be the unknown resistance. When the battery key and the galvanometer key are depressed (the battery key a little in advance), a current will in general flow through the galvanometer, giving rise to a deflection. But if the four resistances fulfil the condition $\frac{P}{Q} = \frac{R}{S}$, no current passes through the galvanometer. Adjust, then, by whatever means are provided, the resistances, till on pressing the keys no deflection takes place. In this and many other cases it is necessary practically to adjust first, so as to obtain a small deflection one way, and then to obtain one the other way. An intermediate adjustment will then give the true condition for no deflection. If this precaution were not taken, a bad contact somewhere in the circuit might be responsible for the absence of deflection. Moreover, this method enables one to estimate the sensitiveness of the arrangement. A common cause of "no deflection" is a break in a cotton covered connecting wire. This is often very difficult to detect when the wire is connected up to the apparatus, on account of the cotton holding the two parts together, but out of actual metallic contact.

I. The Metre Bridge

Here P and Q form a continuous wire resistance. The galvanometer key takes the form of a contact piece, such as a small plate of copper attached to a wooden handle and connected to the galvanometer wire. The ratio $\frac{P}{Q}$ is varied by touching the bridge wire at different points with this contact piece. The resistances R and S, R being unknown and S a variable standard, are connected to each other and to the ends of $P\,Q$ by thick copper strips of negligible resistance, fastened to a base, to which also is fixed a metre stick along which $P\,Q$ is stretched, to enable the ratio $\frac{P}{Q}$ to be determined. It is, of course, equal to the ratio of the lengths of the two portions forming P and Q. Whatever the value of S, a point can be found somewhere on the wire PQ such that, when contact is made there, no deflection is obtained. Find this point roughly, and then calculate approximately the value of R, that is, of $\frac{P}{Q}S$. If P and Q be very unequal, alter S so as to make it approximately equal to R, when the balance point will be not far from the middle of PQ. Then make an exact determination of R. Interchange R and S and repeat. Take the mean value for R. Do this for each of the wires to be measured. Since S is known in ohms, the final results will be given in ohms. It is not necessary to know the resistances of P and Q, as only the ratio of these is required, and this, as we have seen, is equal to the ratio of the lengths.

II. The Post Office Box

Check the results by use of the Post Office Box. This is a Wheatstone bridge complete with tapping keys for battery and galvanometer. P and Q are called the ratio arms of the bridge. Several different ratios $\frac{P}{Q}$ are available. Using any of these, preferably beginning with $P = Q$, the standard resistance arm, also enclosed in the box, can be varied till a balance or an approximate balance is found. (If, however, the value of the unknown resistance be outside the range of the standard, another ratio of P to Q must be used.) Suppose with $P = Q$ and $S = 20$ ohms a deflection to the left is obtained, and with $S = 21$ ohms a deflection to the right. The unknown lies between 20 and 21 ohms. Now alter the ratio $\frac{P}{Q}$ to $\frac{1}{10}$. Deflection to left is obtained with, say, $S=203$ ohms, and to the right with $S = 204$ ohms. R lies between 20.3 and 20.4 ohms. Again, making $\frac{P}{Q} = \frac{1}{100}$, deflection to left is obtained with $S = 2036$ and to right with $S = 2037$ ohms. R lies between 20.36 and 20.37 ohms. It may be possible to form an idea of the next figure by observing the magnitudes of the two deflections. In most P.O. boxes R is the standard and S the unknown resistance. The actual box must be compared with the general diagram, Fig. 90, and fitted up accordingly. Often letters are stamped near the various terminals to indicate the connexions to be used. These may correspond to Fig. 90, or to an arrangement in which battery and galvanometer are interchanged. Either method may be followed.

Specific Resistance

The resistance of a wire depends on its length, sectional area, and nature of material. It is given by $R = \frac{\rho l}{a}$, where l = length, a = area of section, and ρ is called the specific resistance of the material. It is the resistance per unit length of a wire whose sectional area is unity. Measure l and d (= diameter—measure this with a screw gauge) of each wire whose resistance has been found, and calculate the specific resistance of the material of which it is made. Similar wires of the different substances would have resistances in the ratio of their specific resistances. Choose the material whose ρ is least, and obtain some cotton or silk covered wire about 1 mm. diameter of this material. Measure its diameter and calculate the resistance of one meter of the wire.

Construction of Simple Galvanometer

Obtain a wooden ring about 20 cms. diameter, having a groove around it in which a coil of 10 turns of the wire, side by side, can be wound. It should be deep enough to hold two layers of wire. The section of the ring is shown in Fig. 91. If this be not available, build up some kind of substitute. Wind carefully a layer of 10 turns and pass the ends through holes in the rim—a convenient method is shown in the figure. Determine the diameter of the coil. This can be obtained, if desired, before winding, by noting the exact length of a piece of the wire necessary to wrap around the ring and dividing by π. Cover the coil with a layer of stiff paper and wind two coils, of 5 and 2 turns respectively, side by side on the paper. Lead out the ends as before and determine the diameter of the coils. Cover with tape. Prepare a stand to hold the ring vertically, and also to hold the magnetometer, minus its graduated rod, with the centre of its needle at the centre of the coils. The arrangement is shown in the figure. Binding screws may be fixed in the base and connected to the coil ends. The whole forms a simple tangent galvanometer. Calculate the resistances of the coils and, if

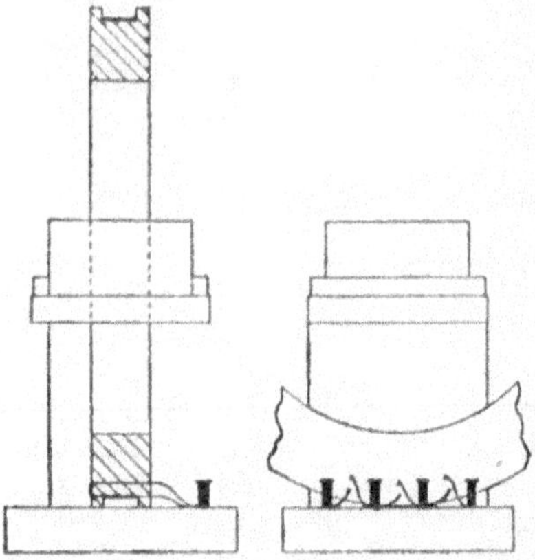

FIG. 91.

not negligibly small, measure them. At least, test by means of the metre bridge to ensure that no appreciable resistance has been introduced in connecting the wires to the binding screws. Take care that no iron nails or magnetic substances are used in any part of the apparatus.

MEASUREMENT OF CURRENT

Adjust the galvanometer so that the needle lies as accurately as possible in the plane of the coil, i.e., till the plane of the coil is in the magnetic meridian. When taking the value of a deflection θ due to the passage of a current, four readings are necessary—both ends of the pointer must be read, and the current reversed and both ends read again. In this way errors due to inexact centring of pivot and circular scale of magnetometer, and to inaccurate setting in the magnetic meridian, are eliminated, provided that these inaccuracies are small.

Connect the 5-turn coil to a copper voltameter, a variable resistance or rheostat, and a 2-volt accumulator, interposing a reversing key between the galvanometer and the rest of the apparatus, as in Fig. 92. A second key somewhere in the

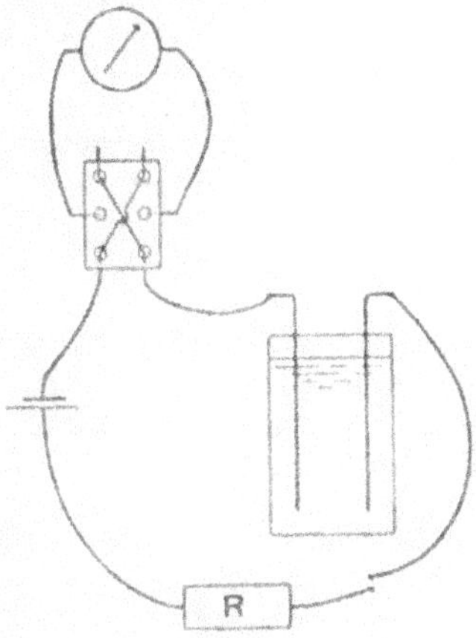

FIG. 92.

circuit is desirable, unless the reversing key will also act as a cut off.

The reversing key is used for the purpose of reversing the current in the galvanometer only, in order to enable deflections on both sides of the zero to be read. It does not change the current through the voltameter. The latter consists of a glass trough containing a saturated solution of copper sulphate, with two copper plates or electrodes suitably supported in it. First perform a rough trial experiment. Wash, dry, and weigh the copper plate which is to act as cathode. Place it in position in the solution, and close the circuit. Note the time, and quickly adjust the rheostat so as to obtain a galvanometer deflection of about 45°. Let the current pass for say 10 minutes, cut off, and take out the cathode. Wash carefully in distilled water, dry high up over a Bunsen flame and weigh. The mass of copper deposited will thus be obtained. In the final experiment a longer time, say 40 minutes, may be allowed. Still, the amount deposited may be too small for accurate measurement. In this case a larger current must be used, and, as a deflection much greater than 45° should not be allowed, it will be necessary to use the 2-turn coil of the galvanometer instead of the 5-turn coil. For a given deflection the current will be inversely as the number of turns, so an estimate can be made of the amount of deposit to be expected whichever coil be used. Set up the apparatus and adjust the rheostat so as to give the required current.

Next, perform the actual experiment. Clean the cathode with nitric acid, rinse with tap water and distilled water, dry and weigh. Replace the cathode and at a given instant close the circuit. Allow the current to flow for, say, 20 minutes. Note the deflection from time to time. If necessary adjust slightly the rheostat to keep this constant. Or else note the deflection each minute and take the mean if slight alterations occur. Reverse the current in the galvanometer and continue the experiment for another 20 minutes. Then cut off the current, take out the cathode, rinse carefully in distilled water, dry and weigh. Assume the value of the electro-chemical equivalent of copper, that is, the number of grms. deposited by one ampere in one second, and hence deduce the value of the current in amperes.

The law of the tangent galvometer is :

$$\text{Current (in amperes)} = k \tan \theta,$$

where k is a constant. Calculate the value of k for the coil used. Deduce the value of k for each of the other coils. Thus if k_2, k_5, k_{10} represent the values for the three coils, we have

$$k_2 : k_5 : k_{10} = \tfrac{1}{2} : \tfrac{1}{5} : \tfrac{1}{10}.$$

In future the galvanometer can be used as a measurer of current. The value of k depends on H, which we have assumed to remain constant.

Determination of Electro-Chemical Equivalent of Copper

It is instructive to use the observations which have already been made in another way. We saw in Chapter VIII that the force at the centre of a circular coil of n turns, of radius r, carrying a current of C electro-magnetic units, is equal to $\frac{2\pi n C}{r}$ in dynes on one unit pole. If C be reckoned in amperes, the force will be given by

$$F = \frac{2\pi n C}{10 r}. \quad (C \text{ in amperes}).$$

This is the force at the centre of our galvanometer coil. To connect this with the deflection observed, we have

$$F = H \tan \theta,$$

as we have already seen under magnetism, and therefore

$$\frac{2\pi n C}{10 r} = H \tan \theta,$$

or

$$C = \frac{10 r H}{2\pi n} \tan \theta \text{ (in amperes)}.$$

But we have already determined the value of H; and r and n are known, therefore C can be calculated.

In making this determination of C we have not relied on anything but fundamental theory and our own experiments. Check the value—and this amounts to testing the accuracy of our determination of H among other things—by working out the value of the electro-chemical equivalent of copper, i.e., by dividing the mass of the deposit of copper by C and by the time in seconds—and compare with the standard value. It is evident, incidentally, that

$$k = \frac{10 r H}{2\pi n}.$$

Calculate its value for each of the coils from this formula, and compare with the values previously determined. [The $\frac{10}{2}$ is kept in, instead of 5, in order to emphasize the relation between amperes and electro-magnetic units. Cross out the 10 and the expression is correct for E.M.U's.]

The values obtained for C by the two methods detailed above may also be compared. They will, of course, be equal if the values for E.C.E. of copper be equal.

Comparison of Forces

The force due to the current at the centre of the coil may be compared with that due to the magnet previously used, at a point on its (the magnet's) axis produced. Choose by reference to Fig. 89 a value for F at a point not too near the magnet. Calculate the current required in one of the galvanometer coils to give a force, equal to this, at its centre. Connect a cell and rheostat with the galvanometer, and adjust till the required current is obtained. This can be done, since $C = k \tan \theta$, where k is known. Now cut off the current, and set up the magnet due E. or W. of the centre of the galvanometer, so that centre of needle lies on the axis of the magnet, produced, at the distance corresponding to the force F. The deflection produced should equal that previously obtained by means of the current, thus showing that the force due to the magnet is equal to that due to the current.

Electromotive Force. I.

To lead up to this, first verify the relation

$$R \tan \theta = \textit{constant},$$

where R represents the resistance of a circuit comprising a cell of constant E.M.F., a resistance box, and the tangent galvanometer. As usual, this must be used in conjunction with a reversing key, otherwise reliable values of θ cannot be obtained. R of course includes the resistance of the cell, which must therefore either be known or be negligible. For present purposes a medium-sized accumulator will give a sufficiently steady E.M.F., and have a resistance small enough to be considered negligible in comparison with the rest of the circuit. Fairly stout and short copper leads should be used, and all contacts carefully made, so as to avoid introducing any appreciable resistance. The resistance of the galvanometer is supposed to be negligible, or else to be known and added to that of the adjustable resistance box. The value of $R \tan \theta$ should be determined for several values of R and its constancy verified. This constancy is due to the fact that $C = \frac{E}{R}$, while C also $= k \tan \theta$, and E and k are constant. We assume that one coil of the galvanometer will be used throughout the present section.

Now plot a curve having C (in amperes) for ordinates and θ for abscissæ, by means of the relation $C = k \tan \theta$, k being the appropriate constant for the coil used in the experiment. This curve, of course, does not specially relate to the experiment, being plotted from the formula merely. Next plot values of R for different values of θ from the above experiment. Now, since $C = \frac{E}{R}$, where E is the E.M.F. in volts of the cell used, at present assumed constant but unknown in value, we have

$$CR = E = constant.$$

Find from the two graphs a series of values of C and of R for a number of values of θ, and multiply the corresponding pairs together. Thus we have, say, $C_1 R_1$ for θ_1, $C_2 R_2$ for θ_2, and so on, giving a number of values for E which should be equal. If slight differences occur, take the mean value for E. Check by means of a voltmeter.

Resistance of a Cell

Assuming $R \tan \theta = constant$ to be verified, we may proceed to determine the resistance of a Daniell cell. Arrange galvanometer, with reversing key, Daniell cell, and resistance box in series. Let B denote resistance of cell and G that of galvanometer, if appreciable. Neglect the resistance of the leads and key. Let R_1 be the resistance used in the resistance box, and θ_1 the resulting deflection, which should be fairly large, say about 50°. Then

$$\frac{E}{R_1 + B + G} = k \tan \theta_1.$$

Increase R_1 to R_2 so as to give a deflection of about half the above value, say θ_2; then

$$\frac{E}{R_2 + B + G} = k \tan \theta_2.$$

By division eliminate E and k. The experiment involves the supposition that E, the E.M.F. of the cell, remains constant, or it could not be eliminated in this way. B can then be calculated.

Repeat for a Leclanché cell. The experiment is not quite so satisfactory, because E varies slightly. To minimize this, take the observations as quickly as possible, so that the current flows for as short a time as possible. Also repeat the first part of the experiment, and take the mean θ for this repeated part and the actual first part. Thus let $R_1\ \theta_1$ be values for first part, $R_2\ \theta_2$ for second, and $R_1\ \theta_3$ for first part repeated. Now, since E has been gradually changing we may suppose that E for first part is rather higher than for second part, while E for third part is rather lower than for second part. Similarly for the values of θ obtained, i.e., θ_1 is rather too high for correct comparison with θ_2, while θ_3 is rather too low. Take, then, the mean of θ_1 and θ_3, and use this in the first of the two equations. Use θ_2 in the second equation. This principle is often useful in the case of quantities which gradually vary.

Electromotive Force. II.

Determine the E.M.F. of the Daniell cell by means of the tangent galvanometer, applying the relation $\frac{E}{R} = k \tan \theta$, where R denotes the resistance of the whole circuit, including resistance box, cell, and galvanometer. Use several different resistances from the box and take the mean value for E. Repeat for Leclanché cell.

Potentiometer Method. Compare the E.M.F.'s of Daniell and Leclanché by the potentiometer method. The apparatus consists of a wire of fairly high resistance stretched along a scale and provided with binding screws at each end, an accumulator with key, and a table galvanometer with contact piece such as was used for the metre bridge, and a three-way key, together with the cells to be compared. The voltage or E.M.F. of the accumulator must be greater than that of either of the other cells. Connect all three + (or —) poles to one end of the stretched wire, and arrange the galvanometer and keys as shown in Fig. 93. To perform the experiment,

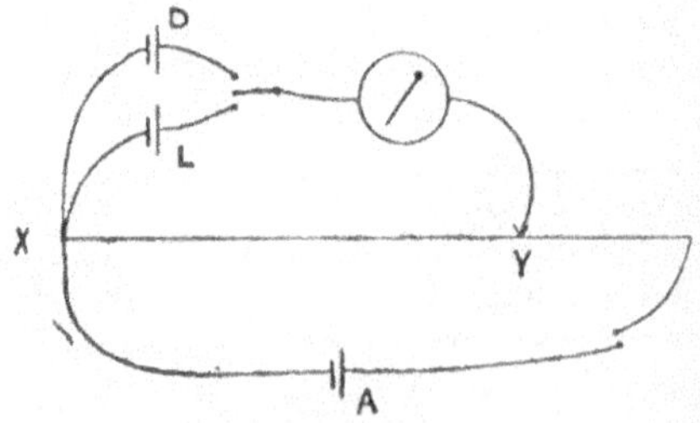

Fig. 93.

close the accumulator key and connect galvanometer to one of the cells by means of the three-way key. Find a point on the stretched wire such that when contact is made there by the contact piece no deflection of the galvanometer takes place. See that opposite deflections can be obtained by moving the contact piece slightly to both sides of the point. Switch over to the other cell, and find another " null point " on the wire. The ratio of the distances of these from the end of the wire to which the cells are joined is equal to the ratio of the E.M.F's. of the cells.

The result depends on the fact that a constant current flows along the wire due to the accumulator A (whose E.M.F. must be constant). The introduction of the loop containing cell and galvanometer does not affect this current, since no current flows along this loop when adjustment is made. There is a uniform fall of potential along the wire, and, if the ends of the loop be X and Y, the potential difference between these is proportional to the distance between them. If, then, no current is to flow in the loop, it is necessary for this potential difference between its ends to be counteracted by the E.M.F. of the cell in the loop. Thus E.M.F. of cell = potential difference XY. The actual value of this is not known, but the ratio for the two cells is the ratio of the two lengths of XY. If the actual E.M.F. of one of the cells be known—that of a properly prepared Daniell is about 1.1 volts—that of the other can of course be at once deduced.

Reduction Factor

In the last experiment a table galvanometer was used, and the method was a " null " one, i.e., the adjustments were made till no deflection was obtained. So long as the galvanometer is sufficiently sensitive to indicate minute currents which flow when the adjustment is very slightly wrong, it does not matter what the law of the galvanometer may be. For small deflections C will be approximately proportional to θ or say $C = k\,\theta$, the constant k being called the reduction factor of the instrument. A very sensitive galvanometer, if used in the last experiment, would need a high resistance in series or a low resistance in parallel with it during the preliminary attempts to find the null point, otherwise it might be broken by the passage of too great a current. In the present experiment a high resistance in series is necessary from a theoretical point of view.

Set up the potentiometer as before, but replace the

Leclanché cell by a plain wire, and put a high resistance in the galvanometer loop. The arrangement is shown in Fig. 94. Find the null point for the Daniell as before. Since no current flows in the loop, the introduction of the high resistance makes no difference, except in that it reduces the sensitiveness of the apparatus to small changes of position of the contact piece. Now switch over to the plain wire. A current flows along the loop, causing a deflection of the galvanometer, but, since the resistance of the loop is much greater than that of the wire XY, the greater part still flows along that wire.

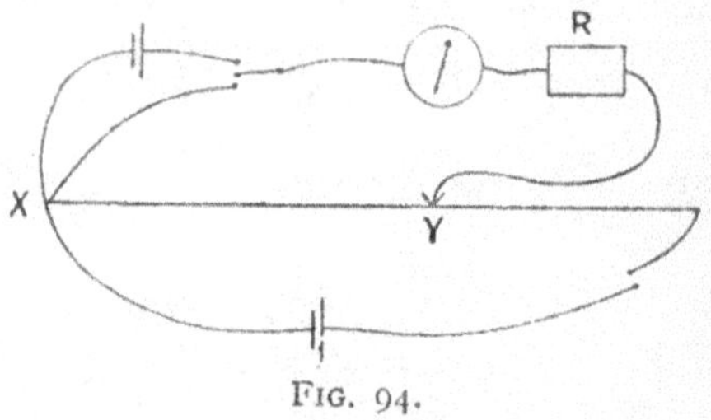

FIG. 94.

We must assume that the flow around the loop is so small as not to interfere appreciably with the constancy of the fall of potential along the potentiometer wire. As in the last case, then (but not quite so accurately), we may say that potential difference between X and Y is proportional to the length XY wherever Y may be. Let XY for the null point determination, when the Daniell was in the loop, be denoted by L, and let the contact piece be put down at various points Y_1 Y_2 Y_3 at distances l_1 l_2 l_3 from X, when the plain wire is in the loop. Let E denote the E.M.F. of the Daniell, and e_1 e_2 e_3 the potential differences XY_1, XY_2, XY_3. Then, since the potential differences are proportional to lengths from X, we have :

$$E : e_1 : e_2 : e_3 = L : l_1 : l_2 : l_3$$

or

$$\frac{E}{L} = \frac{e_1}{l_1} = \frac{e_2}{l_2} = \frac{e_3}{l_3}, \text{ or } e_1 = l_1\frac{E}{L}, e_2 = l_2\frac{E}{L}, e_3 = l_3\frac{E}{L}.$$

Now let the corresponding deflections be denoted by θ_1 θ_2 θ_3, there being of course none corresponding to L. Since

$C = k\theta$, where C is the current in the galvanometer, we may write $\left(\text{since } C \text{ is successively } \frac{e_1}{R}, \frac{e_2}{R}, \frac{e_3}{R}\right)$,

$$\frac{e_1}{R} = k\,\theta_1, \quad \frac{e_2}{R} = k\theta_2, \quad \frac{e_3}{R} = k\theta_3,$$

where R represents the constant resistance of the loop, which is practically that of the resistance used from the box. Thus

$$Rk = \frac{e_1}{\theta_1} = \frac{e_2}{\theta_2} = \frac{e_3}{\theta_3}.$$

But $e_1 = l_1 \frac{E}{L}$, etc., as shown above.

Therefore

$$Rk = \frac{E}{L}\frac{l_1}{\theta_1} = \frac{E}{L}\frac{l_2}{\theta_2} = \frac{E}{L}\frac{l_3}{\theta_3},$$

showing that $\frac{l}{\theta}$ is a constant. Take the mean of the experimental values of $\frac{l}{\theta}$ and put it in the equation :

$$Rk = \frac{E}{L}\left(\frac{l}{\theta}\right).$$

The value of k can be determined, since all the quantities R, E, L, and $\frac{l}{\theta}$ are known.

Temperature Coefficient of Resistance.

Take two samples of wire of metals whose specific resistances have been measured—say, those of highest and lowest specific resistance. Measure lengths whose resistances are exactly 1 ohm in each case, and cut them off, leaving an extra cm. at each end. Solder these ends to stout copper leads. Double the wires in the middle and wind each on a separate cylinder of wood, about 1 cm. diameter, taking care that no contacts occur along the wires. Fit up one of the coils in the steam heater used for specific heat determinations. Measure its resistance accurately by the P.O. box (before passing steam), noting the temperature indicated by the thermometer in the jacket. Now pass steam until the coil is thoroughly and uniformly heated. Note the new temperature and determine the resistance. If the resistance of the leads be not negligible compared with the difference between the two resistances of the coil, measure the resistance of a piece of the wire of which the leads are made, and equal to them in length. Subtract this from each of the other measured resistances. Let R_0, R_{t_1}, R_{t_2} be the resistances of the coil at 0° C., and at t_1° and t_2°, the two observed temperatures. Now for any temperature t° C. we have approximately

$$R_t = R_0 (1 + \alpha t),$$

where α is a constant called the temperature coefficient of the resistance. The values of R correspond to the graph of Fig. 95, where the values of R_0 and R_t are shown as ordinates plotted against temperature. As the graph is a straight line it is only necessary to set up, at the proper points along the axis of temperature, ordinates representing the values R_{t_1} and R_{t_2}, obtained by experiment, and draw the straight line joining their upper extremities. This line produced will cut off from the ordinate at 0° C. a length equal, on the scale adopted for the diagram, to the resistance R_0 of the coil at 0° C. The resistance for any other temperature, not too far removed from the temperatures used in the experiment, can be obtained from the diagram by setting up an ordinate at the required temperature. Find in this way R_{100}.

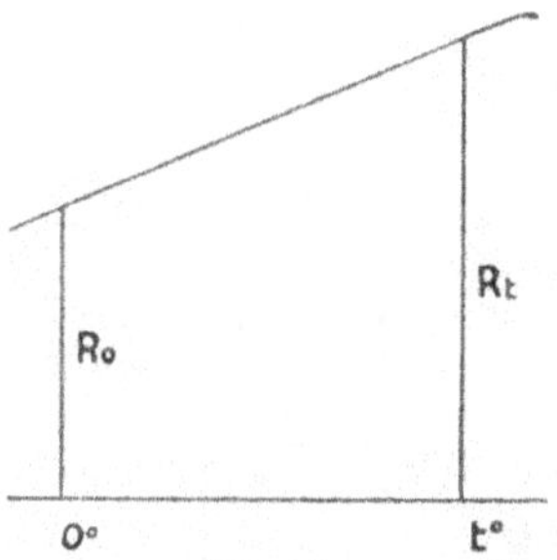

FIG. 95.

Then
$$R_{100} = R_0 (1 + \alpha\, 100),$$

or
$$\alpha = \frac{R_{100} - R_0}{100 R_0}.$$

Repeat for the other wire.

Which wire would be most suitable for use in making a set of standard resistances ?

APPENDICES

APPENDIX A

THE italic symbols, F, l, etc., used throughout this book are meant to represent *physical quantities*. Every *physical quantity* has two aspects. It can be fully specified only by stating (a) its numerical magnitude, and (b) the kind of unit, such as the cm. or the grm., in which its value is to be expressed. This dual nature is expressed when we write "let l = 10 cms." The symbol $[L]$ is commonly used to represent the unit of length, $[M]$ that of mass, and so on. In the text I have used symbols such as **F**, **l**, **m**, etc., to denote the numerical values of forces, lengths, masses, etc. So the *physical quantity* "length" may be fully set out as $\mathrm{l}\,[L]$, or we may write :

$$l = \mathrm{l}\,[L].$$

According to this system it would be incorrect to write "a length l," but correct to write "a length l," or "let l = 10 cms.," or "let l = l cms.," the l in that case simply standing for the number 10.

We have then our *physical quantity* "length" represented for general purposes by l; but when we wish to dissect it we use the form $\mathrm{l}\,[L]$.

Now consider a *physical equation* connecting the *physical quantities* "force," "mass," and "acceleration."

$$F = ma \text{..........................(1)}$$

This, dissected and set out, will be :

$$\mathrm{F}[F] = \mathrm{m}[M]\mathrm{a}[A] \text{...............(2)}$$

or, more fully :

$$\mathrm{F}[M][L][T]^{-2} = \mathrm{m}[M]\mathrm{a}[L][T]^{-2} \text{.........(3)}$$

in which equation the units of force and acceleration are expressed in terms of the fundamental units of length, mass, and time.

Using, however, equation (2), we may decompose into :

the *numerical equation* - $\mathrm{F} = \mathrm{ma}$ ....(4)

and the *dimensional equation* $[F] = [M][A]$......(5)

Thus the *physical equation* (1) or (2) is decomposed into the *numerical equation* (4) and the *dimensional equation* (5).

just as l, a *physical quantity*, was decomposed into l, a *numerical quantity*, and $[L]$, which we may call a *dimensional quantity.*

The equations (1), (4), and (5) might be written in the following way:

$$F = m \times a \quad \text{.......}(6)$$

$$\mathbf{F} = \mathbf{m} \times \mathbf{a} \quad \text{.......}(7)$$

$$[F] = [M] \times [A] \quad \text{.......}(8)$$

In that case we should look upon the sign $\times$ in (6) as a *physical operator*, in (7) as a *numerical operator*, and in (8) as a *dimensional operator*. We need not here pursue the matter further.

APPENDIX B

THE BALANCE

THE practical use of any particular type of balance must be learned in the laboratory. A few general matters, however, may advantageously be dealt with here.

Method of Oscillations. In order to find the "resting point" of an oscillating balance it is not necessary to wait for it to come to rest. Note three successive "turning points," or extreme limits of the swings of the pointer, say to the right, left, and right, on its scale. Take a point halfway between the two right turning points, and then one halfway between this and the left turning point. This marks the place at which the pointer would eventually come to rest.

Determine in this way the resting point for the unloaded balance. Arrest the balance, and put in the left hand pan the object to be weighed. In the other put standard masses; and suppose the least of these available to be 0.01 grm. In a particular case suppose that 20.23 grms. give a deflection of $\mathbf{x}$ scale divisions to the left of the no load resting point, or the zero as we may call it. That is, the new resting point is $\mathbf{x}$ divisions to the left of the zero. Suppose again that 20.22 grms. give a deflection of $\mathbf{y}$ divisions to the right of the zero. Clearly the latter mass is too small; what is required is the addition of a small mass capable of producing a deflection of $\mathbf{y}$ divisions. But the 0.01 grm. just removed produces a deflection of $\mathbf{x} + \mathbf{y}$ divisions; thus the mass required must be $\frac{\mathbf{y}}{\mathbf{x}+\mathbf{y}}$ of 0.01 grm. This value may be worked out and added to the 20.22 grms. In ordinary cases this value should not be carried to more than one place of decimals beyond that given directly by the standard masses used; in this case, for instance, the method should be used only to give the figure in the third decimal place. In the case of balances provided with a "rider" the above operation is not necessary.

Suppose, then, that the correct standard mass has been found which would exactly counterpoise the object to be weighed. It by no means follows that the mass of the object itself has been determined. The first source of error is due to a possible inequality in the lengths of the arms of the balance. Let x = length of left arm, and y that of right arm, and let m_1, m_2 denote the masses in left and right pan respectively, the correct adjustment of the balance having been obtained. The *weights* in the pans will be m_1g and m_2g. Then, by the principle of moments,

$$m_1gx = m_2gy, \text{ or } m_1x = m_2y,$$

so that m_1 will be equal to m_2 only if $x = y$. Absolute equality of x and y is not to be looked for, though of course in a good instrument the difference is very small. The true mass of the object can be found, so far as the matter we are dealing with is concerned, by interchanging the masses in the pans and readjusting the standards so as to obtain a balance. Let this new value of the standards be m_3, the object m_1 now being in the right pan. We have now

$$m_1y = m_3x,$$

and this, by multiplication with

$$m_1x = m_2y,$$

gives

$$m_1^2 xy = m_2m_3xy, \text{ or } m_1 = \sqrt{m_2m_3},$$

showing that the mass required is equal to the square root of the product of the two sets of standard masses used in the two weighings. The ratio of the lengths of the arms can also be obtained by dividing one equation by the other, thus:

$$\frac{m_1y}{m_1x} = \frac{m_3x}{m_2y}, \text{ giving } \frac{y^2}{x^2} = \frac{m_3}{m_2}$$

or

$$\frac{y}{x} = \sqrt{\frac{m_3}{m_2}}.$$

This value for the ratio could be used in future so as to obtain true mass with one weighing only; for

$$m'_1 = \frac{y}{x} m'_2,$$

where m'_1, m'_2 denote the mass of any object in the left hand pan, and the standard mass counterpoising it in the right.

An alternative plan is to place the object in the left pan, counterpoise it with sand, or anything else convenient, arrest, and remove the object, and replace it with standard masses till these balance the sand, using the method of oscillations if necessary. The mass of the object is then equal to that of the standards.

A correction is also necessary on account of the buoyancy of the air in which the weighing takes place. The weight of a body affecting the balance is really the true weight less the weight of the air which the body displaces. Let w_1g, w_2g be the weights (in dynes) of air displaced by the object and by the standards used to balance it. The equation of moments is actually

$$(m_1g - w_1g)\,x = (m_2g - w_2g)y\,;$$

but, since we have already dealt with the inequality of x and y, we will simply write

$$m_1g - w_1g = m'g - w_2g,$$

or

$$m_1 - w_1 = m' - w_2,$$

where m' is the true mass, i.e., $\sqrt{m_2m_3}$ as obtained above. Now the weight of air displaced is equal to the volume displaced $\times$ density of air, and the volume displaced is the volume of the body, or its mass $\div$ density. Let d = density of air, and d_1, d_2 densities of object and standards respectively; then the equation becomes

$$m_1 - \frac{m_1}{d_1}d = m' - \frac{m'}{d_2}d$$

$$\therefore \quad m_1\left(1 - \frac{d}{d_1}\right) = m'\left(1 - \frac{d}{d_2}\right),$$

whence

$$m_1 = m'\frac{\left(1 - \dfrac{d}{d_2}\right)}{\left(1 - \dfrac{d}{d_1}\right)},$$

or with sufficient accuracy, since d is small compared with d_1 and d_2 in ordinary cases,

$$m_1 = m'\left(1 - \frac{d}{d_2}\right)\left(1 + \frac{d}{d_1}\right),$$

or

$$m_1 = m' + m'\left(\frac{1}{d_1} - \frac{1}{d_2}\right)d.$$

The equation is set out in this form so as to show the value of the correction, due to buoyancy, to be added to the mass m' of the standards employed. This correction, $m'\left(\frac{1}{d_1} - \frac{1}{d_2}\right)d$, is positive if the density d_1 of the object be less than that of the standards. The standards are usually of brass, with the exception of the fractional ones; but the air displaced by the latter is so small that no appreciable error is introduced by treating the whole of the standards as if they were made of brass of uniform density. The density of air depends on its pressure and temperature; its value is $.001293 \times \frac{273}{273+t} \times \frac{\mathbf{p}}{76}$ grms. per c.c. for temperature $t°$ C. and pressure $\mathbf{p}$ cms. of mercury, but the value .0012 grms. per c.c. may be taken as sufficiently accurate for general purposes when the weighing is done at ordinary atmospheric temperature. Indeed, in first year work the correction need hardly be made except in the case of bulky objects of small density, or where exceptional accuracy is required.

Sensitiveness of the Balance

The sensitiveness of a balance may be expressed as the number of scale divisions through which the resting point of the pointer moves when a small given mass is added to one of the pans; or, more shortly, as the deflection due to the given mass. Determine the sensitiveness for no load and for various loads up to the maximum for which the balance is used in the following way: Consider the resting point with no load on the pans; add the small mass to one pan, and note the resting point. Transfer the mass to the other pan, and observe the new resting point. Either of the deflections, or, say, the mean of the two deflections, gives the sensitiveness. But it is clear that the original resting point need not be noted, for the same result is obtained by taking half the total deflection from the one extreme resting point to the other. Next add a load to each pan, and observe resting points with the small mass first in one pan and then in the other, and obtain the sensitiveness as before. Repeat for the various loads, and then plot a curve of sensitiveness, showing as ordinates the several values of this, and as abscissæ the corresponding loads.

APPENDIX C

A QUANTITY, like the force acting on a particle, which has, in addition to its actual magnitude, a definite direction in space, is called a Vector. A quantity like mass, which is not concerned with direction, is called a Scalar.

Other vectors are: a portion of a straight line having definite length and direction, the displacement of a particle along a given path in space, velocity (when direction in space as well as speed is implied), momentum, acceleration, etc. Other scalars are: speed (apart from direction), density, temperature, energy, electric potential, electric charge, etc.

A vector may be represented by a straight line, the length and direction of which represent the magnitude and direction of the vector. An arrow head may be added to indicate the sense of the vector.

Suppose a straight line $\overrightarrow{AB}$ to represent a displacement of a particle from A to B, and a line $\overrightarrow{BC}$ another displacement of the particle from B to C. The resultant of the two displacements will be represented by a straight line $\overrightarrow{AC}$. The particle, if it moved along this line, would have the same ultimate displacement with respect to its starting point as if it had traversed the path $\overrightarrow{ABC}$. Now $\overrightarrow{AC}$ is called the sum of the vectors $\overrightarrow{AB}$ and $\overrightarrow{BC}$, and we may write

$$\overrightarrow{AC} = \overrightarrow{AB} + \overrightarrow{BC}.$$

In general, the sum of two vectors S and T, where these vectors can be represented by two straight lines following one another in order, is equal to the vector R, represented by the straight line directed from the beginning of the first to the end of the second of the vectors S, T. It is immaterial whether S or T be taken first.

The same idea applied to forces is represented in the principle of the triangle of forces, which is illustrated by an experiment in Part II of this book. The vector sum of two velocities, say the velocity of a ferry across stream and that of the stream itself (supposed to carry the ferry along with it), gives the resultant velocity of the ferry in both magnitude and direction.

Scalars are added in the ordinary manner of arithmetic.

INDEX

www.ingramcontent.com/pod-product-compliance
Lightning Source LLC
Chambersburg PA
CBHW071418170526
45165CB00001B/321
* 9 7 8 1 4 8 2 0 8 0 3 1 5 *